The Cambridge Manuals of Science and
Literature

WIRELESS TELEGRAPHY

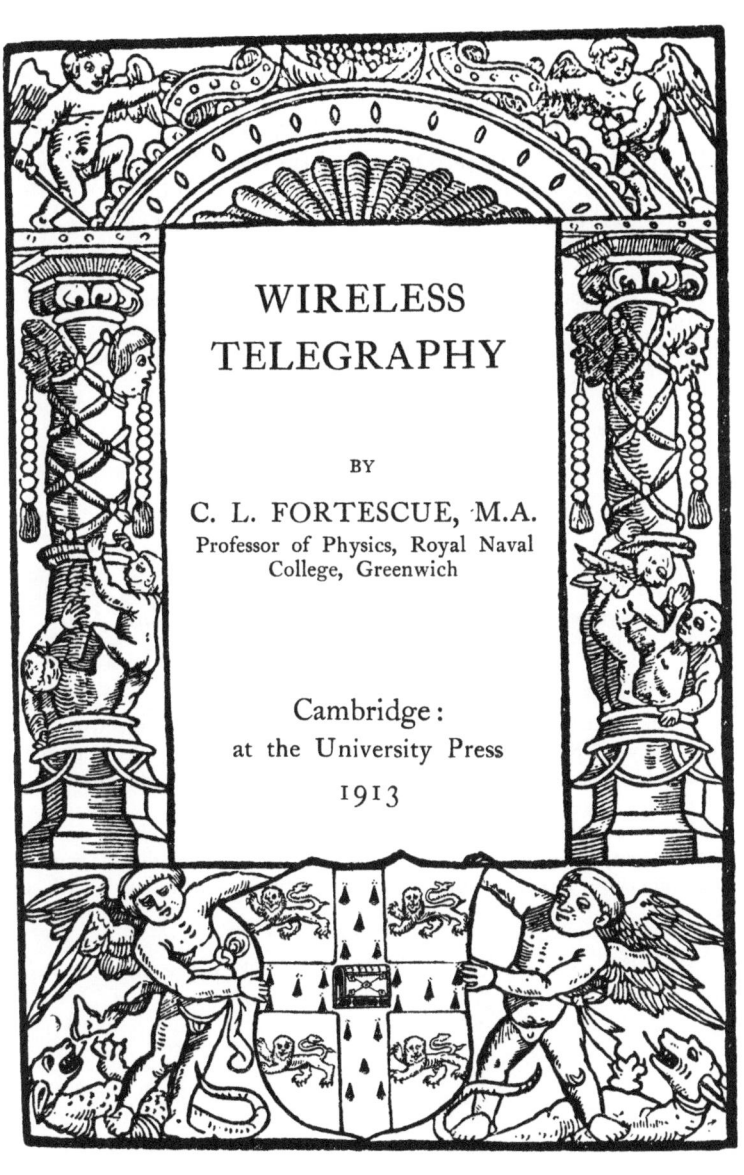

WIRELESS TELEGRAPHY

BY

C. L. FORTESCUE, M.A.

Professor of Physics, Royal Naval
College, Greenwich

Cambridge:
at the University Press

1913

CAMBRIDGE UNIVERSITY PRESS
Cambridge, New York, Melbourne, Madrid, Cape Town,
Singapore, São Paulo, Delhi, Tokyo, Mexico City

Cambridge University Press
The Edinburgh Building, Cambridge CB2 8RU, UK

Published in the United States of America by Cambridge University Press, New York

www.cambridge.org
Information on this title: www.cambridge.org/9781107605909

First published 1913
First paperback edition 2011

A catalogue record for this publication is available from the British Library

ISBN 978-1-107-60590-9 Paperback

With the exception of the coat of arms at the foot, the design on the title page is a
reproduction of one used by the earliest known Cambridge printer, John Siberch, 1521

PREFACE

IN this book the author has had in his mind's eye the reader who, possessing a general scientific knowledge, is anxious to know something, not only of the accomplishments of wireless, but also of the means by which they are attained. The subject is necessarily a highly technical one, and the first four chapters are devoted solely to explanations of the electrical phenomena involved. Which explanations it may be added are really little more than statements of facts illustrated where possible by mechanical analogy. The fifth and sixth chapters deal with the application of these principles to wireless apparatus, and from the seventh chapter onwards the book is devoted to a general survey of the uses to which wireless is nowadays put. No forecast of the future has been attempted as at the time of writing the developement is so rapid that it is beyond anyone to foresee the goal to which it is leading.

A short bibliography has been added including most of the better known books on the subject.

The author's thanks are due to Marconi's Wireless Telegraph Co. Ltd. and Messrs Siemens Bros. and Co. Ltd. for the use of photographs of Marconi and Telefunken apparatus ; and to Mr H. W. Gregson for reading through a large proportion of the proofs and offering many valuable suggestions.

C. L. F.

London.
March 1913.

CONTENTS

LIST OF ILLUSTRATIONS

CHAPTER I

INTRODUCTION

THE general principles on which a message is sent from one place to another by means of Wireless Telegraphy are in many ways analogous to the principles involved in sending a sound signal from one point to another. In the latter case the transmitting instrument, consisting of a bell or syren for instance, is made to set the air surrounding it into a state of rapid vibration. These vibrations spread themselves in all directions in the form of sound waves in the air, and impinging on the receiving instrument, usually the human ear, give rise to effects which enable the sounds to be recognised.

In a Wireless Telegraphy Installation a similar process is taking place during the sending of a message. Corresponding to the mechanical vibrations of the bell or syren are the electrical oscillations in the aerial at the transmitting station, whose function it is to set the aether surrounding it into a state of violent electrical vibration. These electrical vibrations in the form of electromagnetic waves in the aether are transmitted in all directions and impinging on the receiving aerial, can be recognised by means of the receiving instruments.

The apparatus required at the transmitting station to set up these vibrations, and at the receiving stations to recognise them, is, however, rather more complicated than that required to give rise to and recognise the sound waves, when signalling by that means; and consequently before describing these instruments it is proposed to give a brief account of the electrical phenomena upon which their action depends.

The most familiar idea of electricity is the Current of Electricity, generally regarded as the flow of a kind of intangible fluid whose behaviour is similar to the flow of a stream of water along a pipe. Electricity, looked upon in this way, will flow freely through some materials, the most important of which is copper, which are known as "Conductors" of electricity; but only with extreme difficulty through others which are called "Non-conductors" or "Insulators." Well-known materials of the latter kind are air, glass, mica, ebonite, etc. A long copper wire forms a path for electricity to flow along just as the inside of a water-pipe forms a space along which a stream of water can flow. The quantity of electricity passing a point in the wire in, say, a second is analogous to the quantity of water passing a point

in the pipe in a second.   When a large quantity of
electricity passes in a second the current is spoken of
as being a large one, and *vice versa*.   If the wire
along which such a current of electricity is flowing
is broken the current will cease because at the break
a non-conducting barrier has been introduced.   This
is equivalent to putting a valve into a water pipe
and closing it, or to blocking up the pipe with some
impermeable material.

To force water along a horizontal pipe there must
be a difference of pressure between the two ends,
which must be maintained if the stream is to continue.   Similarly in the electrical case, there must be
a difference of electrical pressure to cause electricity
to flow from one end of a wire to the other, which
must be maintained if the current is to continue.   It
may be pointed out that this electrical pressure is
not of the same mechanical nature as the pressure in
the water-pipe.   It cannot be measured in pounds
per square inch or tons per square foot; but it is
nevertheless closely analogous to it.   Electricity cannot
flow without a difference of electrical pressure, but a
difference of pressure can exist without a current if
there is no conducting path for the electricity to flow
along.   Electrical pressure is often called Electromotive Force.

If water is supplied from a pressure main to a
water motor through a small pipe a large proportion

of the pressure is lost and only the remainder is available for working the motor. This is bad engineering as the water is wasting its energy in the pipe in overcoming friction instead of doing useful work in the motor. To prevent this wastage a larger pipe must be used. A similar thing occurs in the electrical case. If a current of electricity is sent along a small wire it wastes its energy, and to prevent this wastage a thick wire must be used. The thin wire is spoken of as having a "high resistance" and the thick one as having a "low resistance." For wireless purposes it is highly undesirable for the current to waste its energy in this way, and so all circuits should be of low resistance, i.e. of large sized copper wire, and where the currents are very large, large copper tubes must be used. Copper is used because size for size the wastage is less with copper than with any other practicable material.

Electricity can under some circumstances be forced through a non-conducting break in a circuit. This occurs with a high electrical pressure and is of the nature of a disruptive spark bursting through the insulating medium. After the initial spark has formed the path through the insulator becomes momentarily a fairly good conductor of comparatively low resistance on account of the volatilisation of the materials. In many of the transmitting instruments used for wireless telegraphy a spark of this kind is continually taking

place between two conductors separated by a layer
of air. This point at which the sparks are formed
is generally known as the "spark gap." With air as
the insulating medium which is burst through, the
non-conducting state quickly restores itself when the
current ceases and the same high electrical pressure
will be necessary to produce another spark.

A stream of water may flow in many ways. For
instance it may flow steadily in one direction at a
uniform rate. If a tap, connected to a constant
pressure main, is left turned on this steady flow will
take place. Again it may flow backwards and forwards, first in one direction and then in the other as
is approximately the case at the mouth of a tidal
harbour. Corresponding to these two ways there are
two ways in which electricity can flow. It can flow
steadily in one direction, in which case it is known
as a Direct or Continuous Current. Or again it can
flow alternately in one direction and then in the
other, when it is known as an Alternating Current.
The variations of an alternating current repeat themselves regularly after certain intervals. Starting
from an instant at which the current is zero, as at 0,
fig. 1, the changes are as follows : at first the current
increases to a maximum value at A and then dies to
zero again at B. During this time the electricity has
been flowing in one direction only, but beyond B the
flow is in the opposite direction, the current being a

maximum at C and falling to zero again at D. This
series of changes is what is termed a "complete
cycle." The two halves of the cycle occupy the same
time and at corresponding instants the currents are
equal but of course in opposite directions. The
"frequency" of an alternating current is the number
of cycles passed through per second.

As well as these two there is a third important
way in which a current of electricity may vary, which

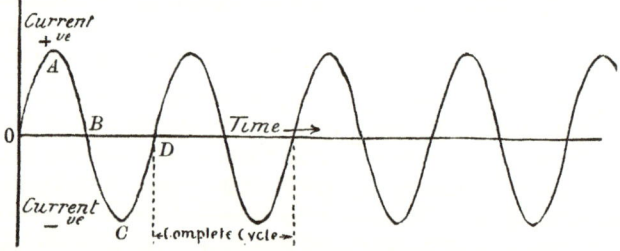

Fig. 1.   Diagram of Alternating Current

is confined almost entirely to wireless work. It may
flow in groups of oscillations to and fro, the amplitude
of the oscillations of each group gradually decreasing
and the intervals between the beginning of each
group and the next being generally large enough for
the oscillations to die down entirely in the meantime.
Such currents are known as Oscillatory Currents. As
an analogy, suppose a U-tube having arms about a
yard in length to be filled with water to within about

a foot of the top. Under normal conditions the level of the water would be the same in each arm, but it could be made different by, say, blowing down one of them. Let the two levels be made different in this or some other way and let the constraint be suddenly removed. The levels immediately tend to readjust themselves and if the bore of the tube is not too small there will be several oscillations up and down before the water finally comes to rest. Suppose the disturbances of the level to be repeated at regular intervals, generally greater than those required for the water to settle down, and the resulting motion of the water is analogous to the oscillatory currents. Each group of oscillations is called a "Train" of oscillations and may consist of one or two only or of a large number of them. The term "cycle" is used with oscillatory currents to denote two consecutive surges to and fro. These are almost similar to a cycle of an alternating current, the only difference being that with the oscillatory current the amplitudes of consecutive half cycles are steadily diminishing instead of remaining constant as with the alternating currents. Fig. 2 is a diagram of an oscillatory current. The term "frequency" is used with oscillatory currents in practically the same sense as with alternating currents. It is the number of cycles which would take place in a second *if the train continued uninterruptedly for that time.* In practice the trains do not last as long as this and there is usually a period of quiescence between them. The actual number of cycles passed through in a second is therefore less than the frequency. For suppose that in a given case there are 100 trains per second each lasting for one-thousandth of a second and consisting of 50 cycles. The actual number of cycles passed through in a second will then be $100 \times 50 = 5000$. But if one train had lasted uninterruptedly for a whole

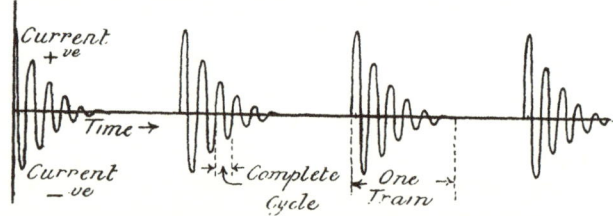

Fig. 2.   Diagram of Oscillatory Current

second the number of cycles passed through would have been $50 \times 1000 = 50,000$. This latter, and not the former, is the frequency of the oscillatory current in question. It is in fact the rate at which the cycles are being passed through during a train, expressed as so many per second.

The frequencies met with in wireless work are for alternating currents 25 to 1000 cycles per second and for oscillating ones 50,000 to 3,000,000.

Each of these three different types of current

involves three corresponding electrical pressures or
electromotive forces.   A direct current will be produced by a steady unidirectional electrical pressure ;
for an alternating one an alternating electromotive
force of the same frequency will be required; and for
an oscillating one an oscillating pressure is necessary.

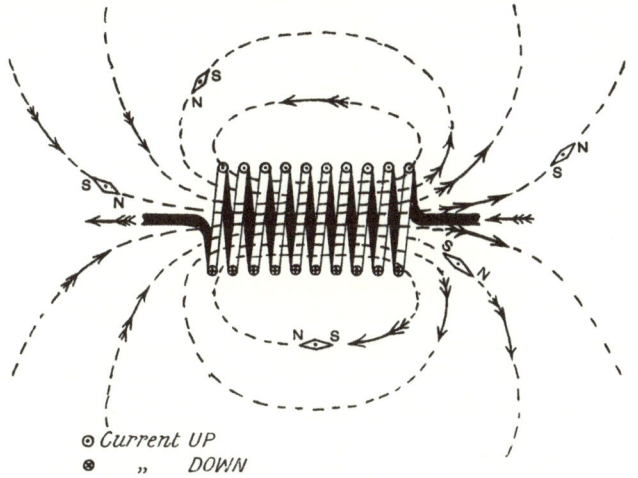

⊙ *Current UP*
⊗   „   *DOWN*

Fig. 3.   Magnetic Field near a coil carrying an Electric Current

When a current of electricity flows along a wire
it produces magnetic effects in the surrounding air.
Suppose for instance that the wire is wound up into
the form of a coil as shewn in fig. 3, and that an
electric current is sent round it.   Then, anywhere in

the neighbourhood of the coil a freely suspended
small magnet will be found to be strongly affected.
Instead of pointing N. and S. as it usually does it will
be found to set itself in directions indicated by the
arrows and lines in the figure. This behaviour of
the small magnet shews the presence of what is
called a "Magnetic Field" around the coil. This
magnetic field depends upon the current for its
existence and varies with it.

A current of electricity flowing round a coil has
further the important property of being able, under
suitable conditions, to give rise to currents in other
coils near it to which it is not in any way connected.
For instance, consider two coils A and B, fig. 4,
A being connected to a battery or other source of
current and B being quite separate from A. Then
whenever the current in A *is made to change* an
electromotive force comes into existence in the coil
B tending to make a current flow round it. If the
two ends of B are joined by a conducting wire
currents will actually flow round the coils; but if
the ends are not joined the electromotive force will
be there and there will be no current. This electromotive force in B is called an "Induced Electromotive
Force," and the current it will give rise to if the
ends of the coil are joined together is called an
"Induced Current." The induced electromotive force
in B is brought about by the agency of the magnetic

field which invariably accompanies the electric current, the existence of which is indicated diagrammatically by the dotted lines in the figure. It is
really the change of this field round about the coil *B*
that gives rise to these effects. Suppose now an
alternating current is flowing round the coil *A*. The

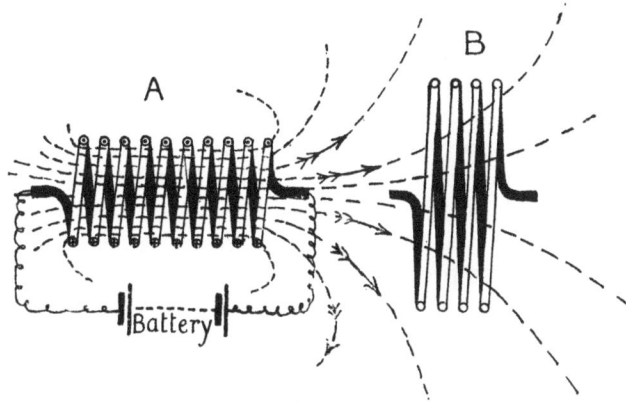

Fig. 4. Mutual Induction between two coils

current and its accompanying magnetic field will be
in a continual state of change with the result that
a varying electromotive force is being continually
induced in *B*. This electromotive force and the
current it can produce are in this case similar alternating ones of the same frequency as the current
in *A*, and may be large or small depending on the

relative position, size, etc., of the two coils. In the
same way if there is an oscillatory current in the
coil *A* there will be induced an oscillatory electromotive force in *B* of the same frequency and dying
away in the same manner. This property of two
circuits is spoken of as their "Mutual Induction."
They are also often spoken of as being "Inductively
Coupled."

Water flowing along a pipe has the property of
Inertia in consequence of which an effort is required
to start it flowing, and when once started gives it a
tendency to go on flowing on its own account. The
sudden closing of a valve at the end of a long pipe
along which water is flowing at a good speed has
been known to burst the pipe near the valve on
account of the pressure set up in bringing the water
to rest. A current of electricity flowing along a wire
has an analogous property, especially if the wire is
wound up into the form of a coil. To start the
current flowing quickly, a large electrical effort (an
electromotive force) is required. Once this current
has been started it will have a tendency to go on
flowing, so much so in some cases that it will burst
its way through a non-conducting break introduced
into the circuit. This is due to what is known as the
"Self Induction" of the circuit and it gives to the
current a property of electrical inertia analogous
to the inertia of water flowing along a pipe. A

circuit in which the effect is very pronounced is said to be "highly inductive," and one in which the effect is negligibly small is said to be "non-inductive." Thus in a highly inductive circuit a large electromotive force will only comparatively gradually set a current going, whereas in a nearly non-inductive circuit a small electromotive force quickly runs up the current to its full value. A coil made up so as to be especially used on account of this property is called an "Inductance." Often also the self induction of a circuit is spoken of as its "Inductance."

The analogy of water flowing along a pipe is of assistance in forming a mental picture of the corresponding electrical phenomena, but there is one very important difference between them which must not be overlooked, and that is that the changes in the electrical case are all very much more rapid than could be the case with the water. A stream of water could be set up in a few seconds, or if a considerable force is exerted, in a fraction of a second. The electrical case corresponding to this is one in which a current is started up in a few millionths of a second only. Similarly the oscillations of water in a U-tube probably require a second or more to complete a cycle, but the analogous electrical oscillation completes a cycle in a few millionths of a second or less.

A Condenser is an instrument in which electricity may be stored, not in the form of chemical energy as in a cell or accumulator, but in the form of actual charges of electricity.

It consists of two conducting bodies separated from one another by any insulating material known as the "Dielectric." Its property of being able to store up electricity is known as its "Capacity." Many condensers consist of a large number of thin plates of metallic foil separated by thin layers of paraffined paper, glass, air or mica. These condensers are known as "parallel plate" condensers. The elevated wires constituting the aerials used for wireless purposes together with the surrounding surface of the earth also form a condenser, the dielectric in this case being the intervening air.

When a condenser has charges of electricity stored in it it is said to be "charged." One conductor may be looked upon as having an "excess of electricity" or positive charge, and the other a "deficiency" or negative charge.

All condensers have this property, namely, that as the amount of electricity stored in them increases the electrical pressure between the two conductors also increases exactly proportionally. If the conducting plates of a condenser are small and some distance apart a large difference of pressure will be required to store up a small charge. Such a condenser is said to have a small capacity.   If the plates are large and near together the condenser will have a large capacity. The amount of electricity which can be stored up in any condenser is limited by the maximum electrical pressure which can exist between the two conducting

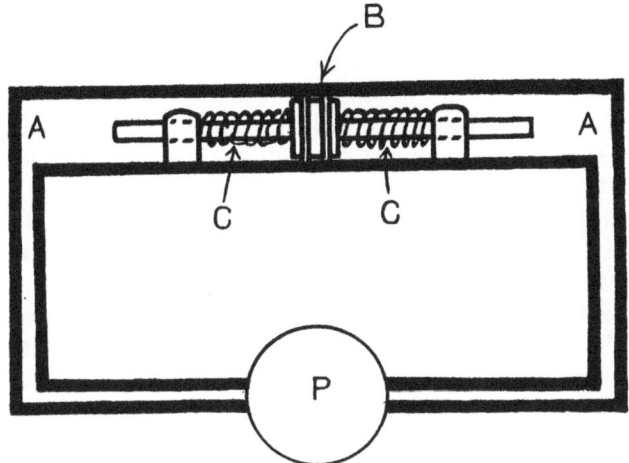

Fig. 5.   Hydraulic model of a Condenser

bodies before a spark jumps across between them, resembling on a small scale a flash of lightning.   This spark consists of an actual flow of electricity, or current of electricity through the dielectric. As a result of this flow the positive and negative charges run together and neutralise one another, the condenser becoming thus "discharged." A condenser can be discharged without damage to the dielectric by simply connecting the two conductors together by a wire. The wire then has a transient current of electricity in it during the time that the charges are running together. These "Discharge Currents," as is explained later, are of the greatest value for wireless telegraphy purposes.

The behaviour of a condenser is well illustrated by the hydraulic arrangement shewn in fig. 5. It consists of a large cylindrical vessel A connected up at each end to a pump P. Sliding in the vessel is a watertight piston B controlled by the springs CC. Under normal conditions this piston divides A into two equal parts. If now the pump is set working water will be drawn out of one side and pumped into the other with the result that the pressure on one side increases and on the other side decreases, the piston being pushed over to the right or to the left. The excess of water on one side of the piston corresponds to the excess of electricity or positive charge on one of the conductors in the electrical condenser ; the deficiency on the other side corresponds to the deficiency of electricity or negative charge on the other conductor. The excess of water on one side over that on the other is proportional to the movement of the piston against the springs, which in turn is proportional to the difference of pressure on the two sides of the piston.

CHAPTER II

OSCILLATORY CURRENTS, RESONANCE AND TUNING

THE importance of condensers for wireless telegraphy lies in the fact that it is principally by their aid that the necessary very high frequency oscillating currents are produced. When a condenser is charged up to a high voltage and is then allowed to discharge through an inductive circuit, the result is a single train of oscillatory currents. The action is as follows: as soon as the inductive circuit is connected up to the two charged plates of the condenser the excess of electricity on one starts running out to the plate on which there is a deficiency, so producing a current of electricity in the wire. The electrical pressure between the plates will quickly speed up this current to a high value in spite of the self induction of the circuit acting like the inertia of a flywheel and opposing the increase of the rate of the flow. A time, however, soon comes when the excess of electricity will have completely run out of one plate of the condenser and it will be for an instant uncharged. At this instant there will be no pressure tending to keep the current flowing and it consequently tends to stop, but owing to the inertia effect of the self induction it will go on flowing with the result that an excess of electricity collects on the plate which originally had a deficiency. The electricity is, as it were, gaining impetus as it flows out from one plate which carries it into the other one in spite of the fact that as it collects there a rising pressure is produced tending to make it flow out again. This rising pressure will eventually stop the flow and start it back in the reverse direction ; the whole process being repeated several times, backwards and forwards.

There are many mechanical analogies to this oscillation of the current when a condenser discharges through an inductive circuit. Take for instance the water analogy suggested in fig. 5. Suppose there to be an excess of water on one side of the piston and the pump to be replaced by a large valve which can be suddenly thrown fully open. Immediately this is done the compressed springs will force the water round from the side on which there is an excess to the side on which there is a deficiency. The speed at which the water flows round will go on increasing so long as the springs are exerting any force. The piston, however, soon reaches its middle position where the springs exert no force. But the stream of water would not stop at this point. Its inertia would carry it on into the side of the cylinder where originally there was a deficiency but where there will now be an excess. As the result of this excess the piston is displaced beyond its middle point and immediately the springs come into action and oppose the further flow of the water, eventually bringing it to rest.   The springs, being compressed, then force out the water in the opposite direction to that of the first stream, the speeding up and over-running being repeated several times.

This oscillation of the water backwards and forwards will go on until the frictional resistance of the pipes brings it to rest, and in point of fact it would not be possible to make an apparatus of this kind which would oscillate to and fro many times. A similar thing occurs in the electrical case.   Energy will be wasted owing to the resistance of the wires and other similar causes in consequence of which the oscillations get less and less, finally dying out altogether.   Unlike the water analogy, however, the electrical oscillations may go on, in some cases, for several hundred cycles before dying out.

This dying away of the oscillations is what is termed "Damping"; and the oscillation is called a "Damped Oscillation."   If it dies away very quickly it is "strongly" damped, but if it dies away very slowly it is said to be "weakly" or "slightly" damped. In some cases it is possible by special means to produce an oscillation of this kind which does not die away.   Such an oscillation is called an "undamped" or "continuous" oscillation, and is often very desirable and much sought after. It is really an alternating current of very high frequency.

These oscillations are illustrated by the curves in fig. 6 which shew the variation of the pressure between the two plates of a condenser and the current flowing into and out of them, during part of an oscillatory discharge. They are represented by the full and broken line respectively, the scales of current and pressure being given.

Oscillations of this kind which are not too strongly damped have the property that the time for one complete oscillation is the same whether the oscillation has a large or small amplitude. This will be noticed to be the case by referring to fig. 6. The frequency of the oscillations is therefore constant whatever their magnitude.

This frequency is known as the "Natural Frequency" of the circuit. It is high or low depending upon the magnitudes of the capacity and the inductance. If both are small the natural frequency is high and vice versa. That this is the case would be expected from considerations of the hydraulic analogy. If the piston has strong controlling springs a small excess of water on one side of the piston over that on the other will give rise to large forces tending to equalise the amounts. This corresponds to a condenser of small capacity in which a small charge causes large electrical pressures. If the

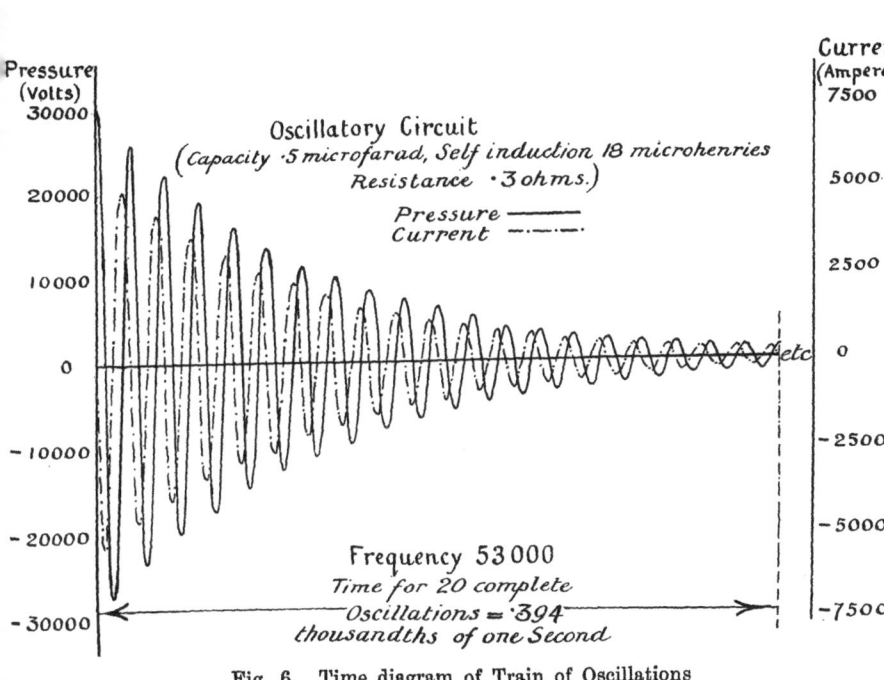

Fig. 6. Time diagram of Train of Oscillations

valve in the connecting pipe is opened the strong springs will quickly force out the excess of water which rushes quickly into the other side and is equally quickly forced out again. Also the shorter the connecting pipes the smaller will be the amount of water the inertia of which will have to be overcome, the more quickly will the strong springs set the water moving and the higher will be the frequency of the oscillations. This corresponds to a discharge circuit of small self induction.

From this analogy it will be easily understood that *small capacity and small self induction mean high frequency, and large capacity and large self induction low frequency.* The frequency as a matter of fact is inversely proportional to the square root of the product (capacity of the condenser) × (self induction of the circuit), and so long as this *product* of these two remains the same the frequency does not in any way depend upon their individual values.

A circuit consisting of a condenser and an inductance with low resistance in which an oscillatory current can thus be set up is generally spoken of as an "Oscillatory Circuit."

If an alternating or oscillating electromotive force be applied to a circuit consisting of a condenser and an inductance *it will produce the greatest effect if its frequency is the same as the natural frequency of the circuit.*

Referring again to fig. 5, suppose that the pump can be made to give small impulses to the water firstly to the left, then to the right, and so on. These small impulses if they are properly timed will produce a large oscillation of the water. The timing of the impulses must exactly correspond to the natural frequency with which the water tends to flow backwards and forwards on its own account. If the timing is incorrect large effects are not produced, because no advantage is being taken of the inertia of the water and the forces exerted by the springs.

Many other examples of this same effect could be quoted. For instance, a company of soldiers marching over a bridge receive the order to "break step," that is to cease marching regularly all in step together. The reason for this is that the regular impulse of the whole company marching in step might coincide with the natural frequency with which the bridge would spring up and down on its own account if once started. Accurate timing of the applied impulses would then produce very serious oscillations in the bridge and possibly the destruction of it.

The regular movement of a few sailors from one side to the other of a large ship can be made to set up a heavy rolling. They must move from side to side exactly as required by the natural frequency with which the ship rolls on her own account. The moving of vastly greater loads will produce negligible effects if the timing is incorrect.

These mechanical illustrations are very closely analogous to the electrical case. If the condenser is charged and allowed to discharge through the inductance, oscillations of current take place of a definite frequency. This corresponds to the natural frequency of the oscillations of the bridge or of the rolling of the ship. The applied electromotive force corresponds to the regularly applied impulses to the bridge or the ship and like them produces the greatest effect when properly timed.

In the mechanical case the applied impulses will have to be reversed in direction after intervals which may be measured in seconds or tenths of seconds. In the electrical case the frequencies are very high and the impulses will have to be reversed after intervals measured in millionths of seconds.

The principle of the timing of small impulses so as to produce large effects is known as the principle of "Resonance." If the frequency of an applied electromotive force is the same as that of the circuit to which it is applied the two are sometimes spoken of as being "in resonance."

Two oscillatory circuits are often so arranged that there is mutual induction between them. In that case when an oscillatory current is set up in one of them a similar oscillatory electromotive force

will be induced in the other. If the two circuits
have different natural frequencies this induced oscillatory electromotive force will not generally produce
a large oscillatory current. When however the frequencies of the two circuits are the same the induced
electromotive force in the second circuit will be of
the right frequency to set up a large oscillatory
current there, and if the trains of oscillations last
long enough the oscillatory current will reach a high
value even though the electromotive force is itself
quite small. It should be remembered that in
speaking of a current being gradually set up in this
way each separate train of oscillations has to be
considered separately. During the intervals between
the trains in the first circuit no electromotive force is
being induced in the second and there is no current
there.

As is pointed out in the next chapter, two oscillatory circuits placed a long way apart may be made to
influence one another by the aid of electromagnetic
waves. The impulses produced in one circuit by the
oscillating currents in the other are then extremely
small. Unless the timing of these impulses is exact,
the impulses cannot be expected to produce any
appreciable result. Properly adjusted, however, quite
an appreciable oscillatory current can be set up in
the distant circuit.

The adjusting of two circuits to the same natural

frequency is known as "Tuning." The adjustment of any circuit is made by varying the self induction or the capacity. In some cases, where a wide range of adjustment has to be provided for, both can be varied.

This tuning of two aerials to the same frequency is of the utmost importance and is the fact upon which the practical utility of wireless telegraphy depends. All over the world there now exist hundreds of wireless stations, some of which are very powerful. Any one of these stations when "sending" will radiate its electromagnetic waves in all directions. As they advance they will strike other aerials and give rise to oscillating electromotive forces in them. The effects of these electromotive forces are, however, generally negligible unless the receiving aerial is tuned to the frequency of the sending aerial. This tuning can be effected by a simple adjustment and it is therefore possible to arrange the receiving aerial to be responsive to the waves from a particular sending aerial or not at will. Thus, if several stations are sending at the same time and using waves of different frequencies any receiving aerial can be adjusted to receive from any one of them and neglect the others as desired. The practical value of this is obvious because otherwise it would not be possible for more than one pair of stations to be working in the same locality at the same time.

It sometimes happens, however, that two aerials whose frequencies are nearly the same are sending at the same time. Each receiving station will then receive its own signals strongly and the others weakly. The state of affairs is then rather like two people talking at once, one loudly and the other softly. A certain amount of confusion is inevitable. The same trouble may arise in another way. If one station is endeavouring to receive from another one at a distance at the same time that a third station near at hand is sending, then the powerful impulses of the adjacent station may set up appreciable oscillations in spite of the frequencies being widely different, and the signals from the distant station may be quite swamped. Confusion of this kind is what is termed "Interference." There are other forms of interference especially noticeable when thunderstorms are near, which are caused by the atmosphere discharging itself to earth through the aerial. If these discharges take place when a message is being received they may obliterate whole letters or whole words. These discharges are what are termed "Atmospherics" and are the bugbear of wireless operators, more particularly in the Tropics.

An aerial which is arranged to respond very freely to its own frequency but only very slightly to impulses differing but little from this natural frequency is spoken of as being very "selective." A high degree of "selectivity" is clearly a most valuable adjunct. A very selective aerial has however the slight disadvantage that its tuning with the sending aerial must be very accurate, otherwise signals intended for it will be passed over.

The nature of the oscillations in the sending aerials will have a large effect on the selectivity. If the oscillations are strongly damped the impulses in the receiving aerials will be few in number, and, in spite of accurate tuning, will not generally have time to set up large currents. If, however, the arrangements are such that these few impulses can set up large currents, then the conditions will also be such that impulses of other frequencies will do the same, and selectivity will be lost. For suppose the damped train of waves to consist practically of one large impulse only. An aerial which can receive this signal will be in a condition in which this single blow sets up the necessary oscillations. But a blow under such circumstances will set up practically the same oscillations whether the sending aerial is tuned to the receiving one or not, and if all sending aerials were to send out these damped waves it would be impossible for more than one pair of stations to work at a time in any given locality. With very slightly damped oscillations in the sending aerials, on the other hand, there will be in the receiving aerials long series of impulses which with correct tuning will gradually set

up the necessary large currents. Impulses of the same
magnitude but of different frequency will not cause
these currents because they are wrongly timed. Thus
with the sustained oscillations a receiving aerial may
be adjusted to pick out messages sent at a particular
frequency and to neglect all others. It is largely for
this reason that as nearly as possible undamped
oscillations are desirable in all transmitting aerials.

CHAPTER III

ELECTROMAGNETIC WAVES

WHEN a condenser of the parallel plate type is
charged up, the dielectric between the conducting
plates is in a condition different from that when
the condenser is uncharged, owing to the electrical
pressure then existing between the two plates.
This electrical pressure gives rise to what is termed
an "Electrical Strain" in the dielectric, or more
probably in the all pervading aether contained in
the space occupied by the dielectric. With the
parallel plate type this strain is confined to the
part of the dielectric immediately between the two
plates, just as when a piece of thin india-rubber is
pressed between two flat surfaces the compression
is distributed over the part of the rubber between

the surfaces only.   If this condenser is connected
up with an inductance so as to form an oscillatory
circuit and if an oscillating current is set up, then
the condenser will be rapidly charged and discharged
first in one direction and then in the other.   Accompanying these rapid charges and discharges there
will be rapid changes in the state of the dielectric
between the conducting plates.   The electrical strain
will be alternately in one direction and then in the
other, but it will be confined to that part of the
dielectric immediately between the plates and will
not tend to spread itself out beyond these limits,
except to a very small extent.

During the time that the oscillatory current is
flowing an oscillatory magnetic field will be produced
around the inductive part of the circuit.   The effects
of this magnetic field will be mostly confined to the
region of the inductance though it will be noticeable
much further away than is the case with the electrical
strain of the dielectric in the condenser.

A circuit such as this, consisting of a parallel plate
type of condenser and an inductance, in which the
effects of the electrical strain, and to a large extent
the magnetic field, are quite local, is termed a
" non-radiative " circuit.

It will be noted that in this oscillatory circuit
the greatest magnetic field is produced when the
current has its maximum value, which occurs at the

instant at which there is no charge in the condenser
(see fig. 6), no pressure between the conducting plates,
and therefore when the electrical strain in the dielectric is zero. The maximum electrical strain on the
other hand occurs when the charge is a maximum,
and the current and with it the magnetic field, zero.
During the oscillations there is thus a continual

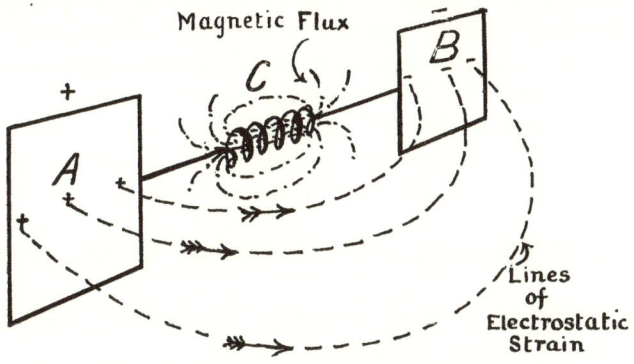

Fig. 7. Radiative Circuit

changing from the maximum of electrical strain to
the maximum of magnetic field and back again. The
electrical strain is alternately in one direction and
then in the other; the same with the magnetic field.

Many circuits in which oscillating currents can be
set up are not thus "non-radiative." Two insulated
plates *A* and *B* (see fig. 7) held horizontally some few

feet apart will constitute a condenser and can be
charged up.   If they are then connected together by
means of an inductive coil C they will discharge, and
an oscillatory current will be set up in the inductance.
In this case when the plates are charged the electrical
strain is no longer confined to a definite part of the
air, but will be distributed somewhat as indicated in
the figure, but of course in all planes round the plates.
The effects will be noticeable at some distance from
the plates and theoretically would extend indefinitely
in all directions.   Now when the condenser discharges
itself through the wire the current will set up a
magnetic field near the coil which again will not be
confined only to the immediate neighbourhood of the
wire.

Such a circuit as this in which both the electrical
strain and the magnetic field spread out into the
surrounding space is termed a "radiative" circuit,
or an electrical "radiator."

Another and more important form of radiative
oscillatory circuit is one in which the conducting
plates consist of a large number of elevated wires
all joined together, for one, and the surrounding
surface of the earth for the other, the intervening
air being the dielectric.   The wires and the earth
are joined by a conductor which may be straight or
may have an inductively wound coil in series with it.
This arrangement constitutes the usual type of aerial

which is an essential part of all wireless telegraphy installations.

With radiative circuits of this kind in which oscillating currents can be set up, the electric and magnetic effects are transmitted to considerable distances by the electromagnetic waves which the oscillating currents give rise to.

It will be well to consider a mechanical case of wave transmission to illustrate these electromagnetic waves. A very simple case in which wave motions can be observed is that of a rope tightly stretched between two supports. If a part of the rope say a few feet from one end is gradually depressed, careful observation will shew that there has only been an appreciable movement of those parts of the rope adjacent to the part depressed. A few feet from that part the rope will apparently be exactly as before. But now if the part of the rope instead of being gradually depressed is suddenly pushed down the same distance by the application of considerable force the results will be quite different. The whole rope will be seen to quiver, a little wave of depression can, with care, be seen to spread in both directions from the part which has been lowered. After a short time the rope will settle down to the same position as when the depression was gradual, but before doing so all parts of it will, on account of the little wave which has run along it, have been moved temporarily through a much greater distance than the final steady displacement, which as previously stated cannot be observed for more than a few feet on either side of the point depressed. If the part of the rope is given a series of say three or four movements up and down in quick succession instead of a single depression only, then a little group of waves can be observed to start from the point of disturbance and move outwards from it. If the end supports are rigid each little group of the waves will be reflected back to the starting-point. As they pass this point the parts of the rope will move up and down just as they were made to do in the first instance but to a smaller extent.   After the group of waves has passed, these parts come to rest and will remain so until a group of waves returns again.   Eventually the rope will come to rest owing to the oscillations gradually frittering away their energy in friction, stirring up the air, etc.

The propagation of the electromagnetic waves is somewhat analogous to the propagation of the little train of ripples along the rope.   The gradual charging up of a condenser of a radiative circuit produces very local effects only.   If the charging up is rapid a zone of electrical strain is set up near the conductors which immediately spreads itself out in all directions with the definite, but very great, velocity of light, namely 186,000 miles per second. This wave of strain produces far greater effects at each point as it reaches it, than the permanent effect produced when the condenser is charged. This is analogous to the way in which the ripple running along the rope causes temporary depressions of each portion of the rope as it reaches it, which are much greater than the permanent ones. (It must be remembered of course that the electric waves spread themselves in all planes and are not confined to one path as in the case of the rope.)

The distribution of the permanent strain in the air surrounding a charged aerial is somewhat as indicated by the fig. 8 a. With an oscillating current in the aerial the elevated wires will be alternately charged in opposite ways positively and negatively. With each charge a wave of electric strain in one direction or the other is sent out from the wires. This is analogous to the ripples produced in the rope when one part is quickly moved up and down several times.

The production of these waves is shewn in a diagrammatic way in figs. 8 a to 8 e, where one cycle of the oscillating current has been taken. Between each positive and negative charging of the elevated wires there will be large rushes of current of very short duration. Each rush of current sets up a magnetic field surrounding the vertical wire in the form of a thick horizontal ring. These rings spread out from the aerial in the same way as the zones of electric strain and in between them.

3—2

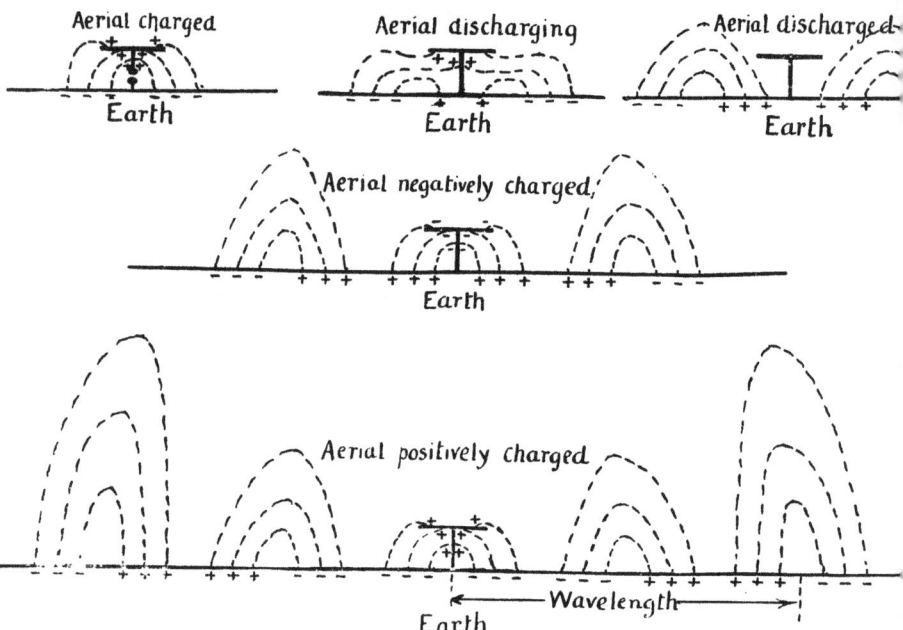

Figs. 8 *a*, *b*, *c*, *d*, *e*. Electromagnetic Waves from an Aerial

The complete waves will therefore be made up of alternate zones of vertical electric strain and horizontal magnetic field. *Each zone expands from the aerial at the rate of* 186.000 *miles per second and there is consequently no catching up or being left behind.*

A "Wave-length" is the distance from a crest to a crest in the case of the wave in the rope. The same term is used for the electromagnetic waves and it means the distance measured in the direction of movement from one state of maximum positive strain to the next one. *The wave-length is consequently the distance a wave can move forward during one complete cycle of the oscillating current producing it.* Thus with an oscillating current of frequency 100,000 cycles per second an interval of $\frac{1}{100000}$th of a second will elapse between the throwing off of consecutive states of positive electric strain. During this interval the first state will have moved forward $186,000 \times \frac{1}{100000}$ miles. The wave-length corresponding to this frequency is therefore 1·86 miles. Waves of lengths from a few hundred feet to four or five miles are used in practice.

An aerial as used for wireless purposes constitutes a condenser of definite magnitude. When it is charged and allowed to discharge, the frequency of the oscillations will therefore depend upon the self induction of the discharge circuit. If this consists of a straight

wire connecting the elevated ones to the surface
of the earth it will have the smallest possible self
induction and the frequency of the discharge will be
as high as can be obtained. If inductances are connected in series with the wire from the aerial to earth
the frequency of the oscillations can be reduced to any
required extent. But since the wave-length depends
upon the time of one oscillation it follows that *by
adding inductance in series between the aerial and
earth the length of the wave radiated can be increased
to any extent.*

Electromagnetic waves of all lengths move with
the velocity of light, and as a matter of fact they are
almost certainly of precisely the same nature as the
wave motion whereby light is conveyed from one
point to the other, the only difference being one of
wave-length. The luminous waves have frequencies
of from four to eight hundred billions of complete
cycles per second and wave-lengths measured in
millionths of a foot, whereas the frequency of the
electromagnetic waves is from 50,000 to a few million
only.

Owing to the very high velocity of propagation of
the electromagnetic waves, the time taken for a signal
to cover a distance of as much as, say, 4000 miles is
very small. The transmission may in fact be looked
upon as instantaneous for practical purposes (other
than considerations of wave length), just as visual

signalling by means of a flashing lamp is usually regarded as instantaneous. For instance, the time required for a signal to cross the Atlantic is about $\frac{1}{80}$th of a second. An observer at the transmitting station can observe the actual spark causing a signal but he would not be conscious of its existence any sooner than an operator at the receiving station 2400 miles away.

The electromagnetic waves carry with them a certain amount of energy drawn in the first case from the oscillating current in the aerial. In consequence of this a radiative circuit is more damped, i.e. its oscillations die away more quickly, than a non-radiative circuit where this dissipation is not taking place. It is also important to note that the rate of radiating energy is much greater for short waves than for long ones, the energy liberated per cycle varying, in fact, inversely as the cube of the wave-length. With an aerial the rate at which the energy is radiated depends also to a great extent upon whether the aerial is used with added inductance in series or not. For a given wave-length an aerial with straight connecting wire gives the maximum rate of radiation, the rate falling off rapidly with a smaller aerial and added inductance.

CHAPTER IV

ABSORPTION AND ATMOSPHERICS

MAXWELL and Hertz in their mathematical theory of electromagnetic waves assume a perfect non-conducting, homogeneous aether ; such an aether as is supposed to occupy interstellar space. The nearest approach to the practical case to which these theories can be applied is to assume the earth to be a flat, perfectly conducting surface and that the surrounding atmosphere does not differ appreciably from the ideal aether. The theory then shews that the intensity of the electromagnetic waves varies inversely as the distance from the transmitting aerial. Thus with oscillations in the transmitting aerial of a given magnitude the electromotive force in the receiving aerial will fall off in proportion to the distance between them. The current will fall off in the same way and consequently the energy in the receiving aerial (depending, as it does, on both the current and the electromotive force) will fall off as the square of the distance between the aerials. But in practice the conditions assumed are not fulfilled, and to determine the extent to which the inverse square law was applicable experiments were carried out by Duddell and Taylor in 1904 between a land station

and a small steamer in the Irish Sea. Up to distances
of about 60 miles the law was found to be approximately correct, but even at that distance the falling
off of the signals appeared to be more rapid than was
given by the theory. This result has been confirmed
since by experiments carried out by the United
States Navy Department over distances up to 1000
miles, which shew that the falling off from the
theoretical value gets greater as the distances
increase.

Practical wireless experience has also brought
to light numerous other facts not explained by the
theory. In the early days when only comparatively
short waves were in use, it was discovered that signals
could be made with more certainty and over longer
distances by night than by day. Later it was found
that by day the short waves had a much longer range
over water than over land but by night the range
was about the same, even though the land was
mountainous or consisted of a sandy desert. Long
waves on the other hand have very nearly the same
range over land as over water whether by day or
night. Mr Marconi, who has probably had more
experience of these phenomena, at any rate over long
ranges, than anyone else, gave some most interesting
facts concerning his long distance transatlantic experiments in a lecture at the Royal Institution on
June 2nd, 1911. Between Clifden and Glace Bay two

waves had been used, one of 7000 metres (23,000 ft.)
and the other 5000 metres (16,400 ft.). The variation
of the strength of the signals on a typical day is
shewn in fig. 9. During the period of daylight at
both stations the signals are approximately uniform
in strength, the long wave being the better one of
the two. In the interval between sunset at the two
stations signals first of all fall off and then rise again

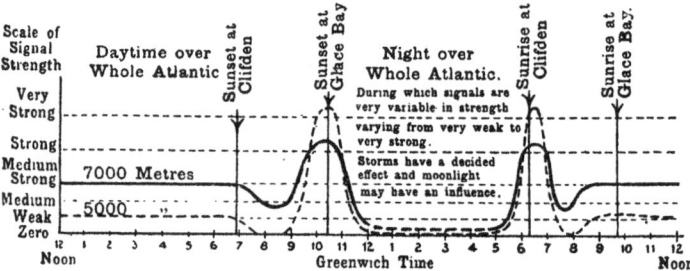

Fig. 9. Diurnal variation of strength of signals at Clifden

to a maximum, the short wave usually being the
strongest. During the night they are uncertain,
sometimes strong and sometimes weak. At the intervals between sunrise at the two stations the same
phenomena are observed as in the intervals between
sunset. Mr Marconi also mentions another remark-
able fact, namely, that it is easier to signal in a
northerly and southerly direction than in an easterly
and westerly one. It had frequently occurred that

ships fitted with the usual installations had been able
to communicate with the English shore stations from
the Mediterranean, distances much greater than could
have been accomplished in an easterly and westerly
direction with the same power.   At Buenos Aires,
Mr Marconi had been able to receive signals from
Clifden and Glace Bay quite clearly by night, over
a distance of 6000 miles.   They could not, however,
be detected by day and yet during this same time
signals between Clifden and Glace Bay themselves
were equally strong by day or night.   The behaviour
in a northerly and southerly direction thus appears
to be quite different from the behaviour in the
easterly and westerly direction.

"Freak" distances have also been numerously
recorded when stations have been able to communicate temporarily over much greater distances
than usual.   It sometimes happens that an operator
can sit at his instruments and hear very distant
stations, the signals waxing and waning in a manner
similar to that in which the sound from a distant
church bell is occasionally noticed to vary on a
stormy day.

The reasons for this very irregular behaviour are,
as previously stated, at present not thoroughly understood.   The falling off of the signals more rapidly
than is indicated by the theory is due to two possible
causes ; either the waves waste their energy *en route*

or they are deflected from the receiving aerial on account of the curvature of the earth or the electrical properties of the atmosphere.

Energy may be wasted in two ways ; owing either to the air being slightly conductive or to the surface of the earth not being a perfect conductor. In the upper strata of the atmosphere a point is reached at which it possibly becomes a partial conductor on account of its rarefaction and by the action of the sunlight. In these layers the waves would be accompanied by conduction currents which would be flowing along paths of high resistance and therefore wasting energy. These layers would become therefore to some extent opaque to the electromagnetic waves just as fog is partially opaque to light waves. This semi-conductive state is thought to be produced by the ultra violet light in the sunlight which brings about a change known as "ionisation." This would explain the fact that short waves have a smaller range by day, when the ionisation is going on most rapidly, than they have by night, but it does not explain the fact that the long waves have practically the same range by day as by night or the fact that the range of the short waves is the same over water as over land at night but differs in the daytime. As the bases of the waves spread out from an aerial, currents of electricity are set up in the surface of the earth. If the surface has a high electrical resistance as would be the case over a sandy desert, these currents lead to a similar waste of energy which would not take place over the sea as the water is a good conductor. This very probably has some connection with the longer ranges which can be obtained over water, especially with the short waves.

The alternative theory is that the waves do not spread out over the spherical surface of the earth in the same way that they would if it were flat. The light waves which are of the same nature, but of much shorter wave length, are certainly propagated only in what are for practical purposes straight lines. From analogy it would not therefore be natural to expect the electromagnetic ones to follow curved paths as they undoubtedly do in spreading out from, say, Clifden to Buenos Aires, a distance of one quarter of the earth's circumference. The problem has been attacked by mathematicians with at present uncertain results, but the trend of opinion seems to be that to some extent waves could thus spread themselves round the surface of the earth, not in the form of space waves, as described on page 35, but in the form of waves on or near the surface of the earth. An alternative possibility is that though in a homogeneous atmosphere the waves can proceed in practically straight lines only, they are reflected and refracted in passing from one layer of the atmosphere to another, thus causing the bending

round the earth which is found to exist in practice.
The electrical properties of the air will undoubtedly
vary slightly with its physical conditions and the
extent of the ionisation.   These variations will cause
slight variations of the velocity with which the waves
are propagated through it, which would cause a
change of the direction in which the waves are
moving.   It is a well-known fact that owing to the
bending of the rays of light as they pass through the
layers of air the whole of the sun's disc is sometimes
visible after it has really sunk below the horizon.
There are reasons for supposing that the depth to
which the ionisation of the atmosphere proceeds
varies with the time of day, being a maximum at any
one place at noon and a minimum at midnight.   Also
it is probable that the extreme outer layers are
permanently ionised to a considerable extent, making
them more effective in bending the waves than the
middle or lower layers.   Further, it can be shewn
theoretically that the long waves are more bent by
the ionised layers than the short ones.   Assuming
the atmosphere to have these properties, it may be
that at night the efficient outer layers refract both
long and short waves back to the earth, even into
the valleys of a hilly country.   But by day it may be
that only the long waves are bent back to the earth
by the less ionised middle layers ; the short waves
being only bent sufficiently to keep them away from

the outer layers and not sufficiently to bring them
back to the earth again before they have frittered
away their energy. A supposition of this kind will
explain the different behaviour of the short and long
waves by day and by night.   It will also more or less
fully explain the variation of the strength of the
signals between Glace Bay and Clifden.   This theory
may seem a little far-fetched in some instances, but
none of the hypotheses involved are wholly devoid
of experimental foundation and it is the only theory
at present in existence which can in any way be
made to explain all the facts.   The suggestion also
offers an explanation of the "freak" distances, the
idea being that a large mass of air may act as an
enormous electromagnetic lens or mirror and focus the
waves on to the particular station when very strong
signals may be received from stations which under
ordinary circumstances would not be heard at all.
There seems so far no way of deciding which, if any,
of these views is the correct one, and until a great
many more experimental data are forthcoming the
uncertainty is likely to continue.

Another cause of uncertainty in the reception of
wireless signals, coming into an entirely different
category from absorption or refraction, is the so-
called atmospheric interference.   This trouble is not
in any way connected with the inefficiency of the
atmosphere as a medium for the transmission of

electromagnetic waves but is due to the powerful
disturbances liberated from flashes of lightning and
other natural electrical discharges. In the case of
a large flash of lightning an enormous amount of
energy is temporarily available, and short but very
violent trains of electromagnetic waves are liberated
which, on striking an aerial, are often sufficient to set
it oscillating in spite of its being tuned to a wave
length very different from that of the waves themselves. The result is that a loud signal is received
sounding like a single "rap" or like a series of two
or more "raps" following one another. When very
bad these sounds continue almost without interruption and render the reception of proper signals very
difficult. Being due to thunderstorms and similar
phenomena they naturally vary in intensity with the
time of year. In the temperate zones they are at
their worst in summer, but are never entirely absent.
The waves are so powerful and have such long ranges
that it is probable that many of those observed in this
country in winter have their origin in the Tropics.
In the Tropics the interference is worse than in
other zones on account of the prevalence of electrical
storms there, especially at certain periods of the
year. During the worst storms the aerial has to be
connected straight to earth to prevent the instruments from being damaged by the heavy discharges.
The same thing has to be done in the temperate

zones, too, if a storm passes directly over the station.

A vast amount of ingenuity has been expended in trying to overcome the interference produced by these stray waves. The use of a high pitched musical note is one of the most effective means of reducing it. The atmospherics, fortunately, are quite irregular and signals in a clear piercing note can easily be read through them if they are not too strong. The strength of all but the very worst can be reduced by the use of very selective receiving circuits, unless the wave length of the atmospheric happens to be the same as that of the wave being received. The very powerful ones, however, cannot be got rid of in this way and are particularly harmful in that they often render the operator's ear insensitive for several seconds, with the result that whole letters or even words may be missed. To prevent this occurring the Marconi Company have devised a very ingenious arrangement of two rectifying detectors acting in the same circuit in opposite directions. One of these is in a sensitive condition but the other is not. For ordinary signals the insensitive one has practically no effect and they are received on the sensitive one in the ordinary way. Very strong signals, however, will bring both of them into action in opposite directions, with the result that the sounds though loud are not deafening. Other arrangements having ultimately the same results are also employed. During the discussion of a paper on this subject before the Institution of Electrical Engineers Mr Duddell suggested the use of more power as the remedy for these troubles. As the available power of some of the recently projected stations is approaching 10,000 horse-power it would appear that his advice was being acted upon.

CHAPTER V

THE TRANSMITTING INSTRUMENTS

THE International Morse Code of "dot" and "dash" signals is almost universally employed for wireless telegraphy just as for telegraphy with wires. It consists of a number of signals of long and short duration, a "long" being equal to two "shorts." By various combinations of these "longs" and "shorts" or "dots" and "dashes" all the letters of the alphabet are represented. To an expert operator the Morse code is as easy to "read" either from photographic records or by means of sounds in a telephone receiver as ordinary print or ordinary conversation. On account of each word being spelt out with letters of from one to four separate signals, the transmission is necessarily slower than speaking or writing. With hand signalling the speed of "sending" rarely exceeds

20 to 25 words per minute (a word being taken as consisting of, on an average, five letters). Some ten times this speed may be obtained by the use of automatic sending by means of apparatus similar to that used for rapid transmitting with the ordinary telegraphy.

The apparatus used for all systems of wireless telegraphy has certain parts in common. In all cases some form of aerial is essential, which is used to radiate the electric waves when a station is "sending" and to absorb the waves radiating from some other station when it is "receiving." Generally the same aerial is used for both these purposes but in some recent large stations separate aerials are used, one for transmitting the other for receiving.

When a given station is "sending," instruments are necessary to generate in the aerial the wave-producing oscillating currents and having produced them to control their duration, so as to produce "dots" and "dashes" as required. It is largely in the means of producing these currents that the systems differ from one another. Considerable differences also occur in the arrangement of the receiving instruments.

In the earliest wireless telegraphy stations the apparatus was of the simplest, and consisted of a small aerial carried on a mast and connected up to one of a pair of spark balls at its lower end (see

fig. 10). The other spark ball was connected to a
wire running down into the ground. An induction
coil was used to charge up the aerial until a spark
jumped across between the two spark balls. When

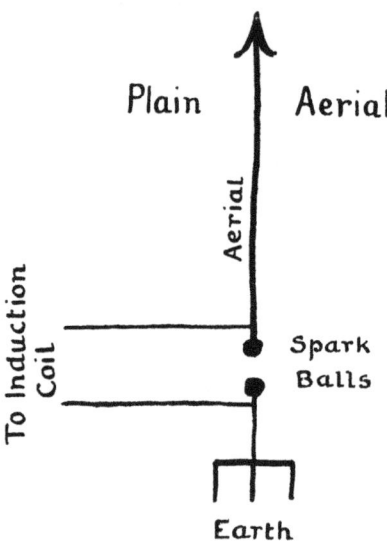

Fig. 10. Diagram of connections for Transmitting by "Plain Aerial"

this took place the aerial discharged itself and an
oscillating current flowed backwards and forwards
from the elevated wire to the earth through the
bridge of heated vapour between the spark balls
which was formed by the first spark. After a few

hundred-thousandths of a second these oscillations ceased, the air resuming its normal insulating condition and the aerial then being ready to receive its next charge from the induction coil. After receiving this charge the process repeated itself, and every time that the aerial discharged a train of electric waves was sent out from it. Each train striking a suitably tuned receiving aerial produced a small effect which was made known in some way or other to the operator on the look out for the message.

If left to itself the induction coil would have gone on charging the transmitting aerial at the fairly rapid rate of some five or ten times per second and at the receiving station there would have been a continuous series of small effects produced. To transmit signals from one station to another it was necessary to interrupt this series so as to divide it up into the necessary "longs" and "shorts" as required for the Morse code. This was done by interrupting the supply of current to the induction coil by means of a hand-operated switch or "key." In this way the discharges could be broken up as required, a "long" lasting for something of the order of one quarter of a second during which the aerial discharged itself two or three times and a "short" for one half of this time, the interval between the signals constituting one letter being of the same length as a "short." At the receiving aerial the small effects would be broken up into exactly corresponding "longs" and "shorts" and the message could be "read."

This apparatus was very simple but had many very serious disadvantages. Its range was limited and it was impossible to render the system selective. This was due to the strong damping, caused largely by the fact that a very large proportion of the energy

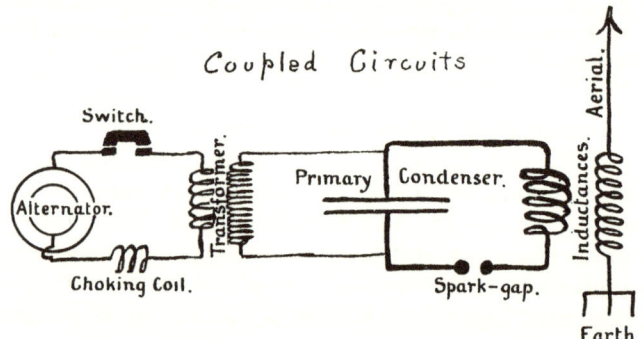

Fig. 11.   Diagram of Coupled Transmitting Circuit

stored up in the aerial just before the spark took place was wasted in the spark gap, and partly, also, by the useful radiation of energy into the surrounding space. The oscillation consisted of one large rush of current followed by two or three very rapidly diminishing ones, and then ceased.

On account of these disadvantages the system was very quickly replaced by the "coupled" system.   In this system there are two quite separate oscillatory circuits *so arranged that they have the same natural frequency* and so that they react upon one another inductively. One of these circuits (see fig. 11) consists of the aerial and the earth connected together by an inductance. The other circuit, often called the primary circuit, consists of parallel plate condensers joined up to an inductance and a pair of spark balls. The inductance in this circuit is so placed that it is inductively coupled with the inductance connecting the aerial to earth. This arrangement thus consists of a non-radiative primary circuit in which oscillations can be set up and which acts as a reservoir from which the radiative aerial circuit can take its supply of energy.

Arrangements are made whereby the parallel plate condenser of the primary circuit can be charged up until a spark jumps across between the spark balls. The condenser then discharges itself through the primary inductance and the spark gap, setting up the required oscillatory current. This oscillatory current in the inductance induces an oscillatory electromotive force in the aerial circuit which in turn gives rise to an oscillatory current there, with the result that electric waves are radiated. With this arrangement a very much longer series of oscillations could be produced in the aerial for one charging-up of the primary condenser. The reaction between the two

circuits is rather complicated, but the final result is
that the aerial radiates a complex wave made up of
two simple waves of different wave-length.   With the
instruments as used in practice the reaction between
the two circuits is not a very powerful one and under
these circumstances the two component waves have
very nearly equal wave-lengths.

Though not ideal this system was a great advance
on the old system of "plain aerial" as it was called.
Ranges were greatly increased and there was an
enormous gain of selectivity owing to the much
longer duration of the oscillatory currents in the
aerial.

The primary condenser is usually charged by
means of an alternating electrical pressure, a spark
taking place generally once for every complete cycle.
This alternating pressure is usually obtained from an
alternator, which is a machine designed to produce
alternating currents, driven by a steam or oil engine
or by an electric motor.   The alternating current is
generated at a comparatively low pressure, lower
than would be suitable for charging the primary
condensers, and is increased to that required by a
static transformer.   This is an instrument having no
moving parts and in which it is easier to secure the
very high insulation necessary for the high pressure.
An inductively wound coil is also included in the
charging circuit so as to give it a natural frequency

equal to that of the alternator. A switch placed in
the charging circuit is generally used for controlling
the time during which the discharges are taking
place, producing thereby the required "longs" and
"shorts." The number of sparks per second when
working depends upon the frequency of the electrical
pressure. Frequencies of from 25 up to 800 and
1000 cycles per second are employed. High frequency sparking is a great advantage where aural
reception is employed. The reason for this is that
each separate spark liberates a single train of waves
which, acting upon the receiving aerial and its detector, produces a single movement of the diaphragm
of a telephone receiver. If these sparks follow one
another regularly and with great rapidity, at the rate
of say 1000 per second, the movements of the telephone diaphragm will be correspondingly rapid and
the sound produced will be a high pitched musical
note. These high pitched notes can be heard when
a lower pitched note would probably be inaudible or
swamped by other noises, particularly the sounds
produced in the telephone by atmospheric discharges
through the aerial. This is just the same as if two
persons were speaking at the same time, one spas-
modically in a low bass voice and the other regularly
in a high pitched piercing treble. A third person
can with a very little concentration follow the latter
without being in any way confused by the former.

There is also the further advantage that the telephone
receiver is more sensitive to high frequency currents
than to low ones.   These high musical notes are
therefore very useful where serious interference is
experienced.   In most modern land stations and in
many of the larger ships' installations, rates of sparking of from 200 to 1000 sparks per second are
employed.   A much greater amount of energy per
second is then, too, sent out from the aerial and
the range is increased.   A corresponding increase
of power is of course required at the transmitting
station.

The disadvantages under which this system labours
are twofold.   Firstly, during practically the whole
time that oscillating currents are flowing in the aerial
there are also oscillating currents in the primary
circuit wasting energy at the spark gap.   Relatively,
the energy wasted in this spark is much less than
that wasted in the spark when using the "plain
aerial" system, but nevertheless the result of this
wastage is that only a part of the energy supplied is
actually radiated in the form of a still somewhat
damped wave train and the spark balls become
so heated that irregular sparking takes place, so
spoiling the clearness of the note in the telephone
at the receiving station.   This is particularly the case
with the rapid sparking previously mentioned, so
much so in fact that all kinds of complicated devices

in the form of powerful blowers, rapidly rotating spark gaps, etc., are being employed, with varying success, to overcome the difficulty.

Of these devices perhaps the most successful is one now used by the Marconi Co., which consists

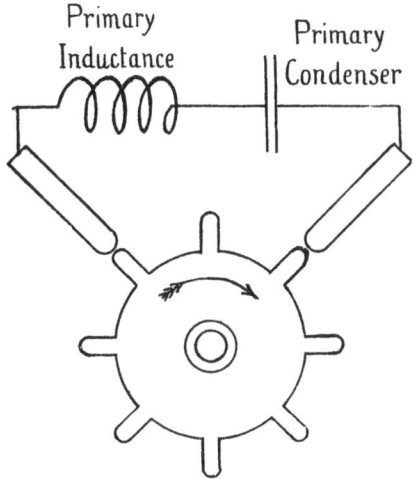

Fig. 12.   Marconi Rotary Spark Gap

of a wheel, on the periphery of which are a number of projecting studs, fixed on to the shaft of the alternator producing the alternating pressure, as shewn diagrammatically in fig. 12. This wheel with its projecting studs rotates between a pair of fixed

studs which are just clear of it. The double gap
from one fixed stud to the wheel and thence to the
other fixed stud here constitutes the gap in the condenser discharge circuit. The position of the fixed
and moving studs is so arranged that they come
opposite to one another just at the instants at which
the spark is required, that is at the instants at which
the condenser is fully charged. When the studs are
not opposite to one another the gap in the discharge
circuit is so large that no spark can possibly take
place. In this way sparks can be obtained with
perfect regularity and a clear musical note produced.
With this device there is no trouble with overheating
of the spark balls on account of the fanning of the
wheel.

With the large Marconi installations an additional
effect is produced with this rotary spark gap. The
peripheral speed of the wheel (which has to be very
high, in the neighbourhood of 36,000 feet per minute)
and the number of studs are so arranged that the studs
are opposite to one another for only a very short time.
Then by the correct adjustment of the coupling of the
two circuits the oscillations in the primary are only
allowed to continue for just long enough to set up
the maximum oscillations in the aerial, the increasing
distance between the studs making it impossible for
them to go on longer. The oscillations in the aerial
then continue whilst the primary condensers are

charging up for the next spark. This final result is almost exactly the same as that obtained by the "quenched spark" system to be explained later.

The second disadvantage of the coupled system is that the waves radiated from the aerial are complex ones made up of two waves of slightly different frequencies, one above and one below the frequency to which each of the circuits are separately tuned. At the receiving station the best results are obtained with the receiving aerial tuned to one or other of these two waves. Signals can, however, be received with the aerial tuned to any intermediate frequency. The result of this is that a wider range of frequency is involved for each "tune" and the whole of the energy radiated is not concentrated in the particular wave for which the receiving aerial is tuned. For the conditions under which this system is used in practice the two waves have very nearly the same frequency so as to overcome this difficulty as much as possible.

Both these difficulties have been largely overcome by the use of one or other of the applications of Prof. Max Wien's "charging by impact" method, notably in the "Telefunken" system and in the nearly identical system of Lepel.

In these systems there are again two coupled circuits as in the system just described. The fundamental difference lies in the fact that in Wien's method means are employed to "quench" the spark in the primary circuit so preventing the current from flowing there for more than two or three cycles. If this is done in a suitable way the whole of the energy of the primary circuit is transferred to the aerial circuit almost in the form of an electrical blow and by the time the primary current has ceased large oscillations will have been set up in the aerial. This circuit then goes on oscillating by itself, the oscillations dying down quite gradually, there being no damping owing to currents flowing across the spark gap. The ultimate result is that a larger proportion of the energy is actually radiated, and this radiation, be it noted, is in the form of a single wave of the natural frequency of the aerial circuit.

The overheating of the spark gap, which with the ordinary spark gap leads to irregular sparking, is not so likely to arise with this method because of the very short time during which the current is flowing there. Also for the same reason the charging up of the primary condenser can start whilst the aerial is still radiating. The system is therefore particularly well adapted to high spark frequencies, and the use of a high musical note which can be "read" through atmospheric interference without difficulty. The oscillations in the aerial circuit are well sustained, making the system very selective and susceptible to fine tuning.

The charging of the primary condenser and the

interruption of the sparking for signalling purposes are brought about in exactly the same way as before.

The same principle can be employed with a direct current for charging the condenser with but slight modifications. This was as a matter of fact the method originally used by Lepel.

The quenching of the spark, which is the distinctive feature of the system, is obtained by breaking the spark gap up into a number of small gaps of about one-hundredth of an inch each between metallic discs of about three inches in diameter. The best adjustment of the circuits and the practical management of the gaps presented some difficulty at first but this has now been entirely overcome.

Besides the two coupled circuit methods just described and generally known as the Marconi and the Telefunken system respectively, there are systems using undamped high frequency currents generated by special types of alternating current dynamos, or as in the Poulsen system by means of an electric arc between suitable electrodes in a non-oxidising atmosphere. In America Alexanderson has designed alternators capable of giving quite large outputs, of the order of twenty to thirty horse-power, at frequencies of 100,000 cycles per second and upwards. Goldschmidt in Germany working on entirely different principles has achieved the same final results. The Telefunken Co. have also recently introduced a method of generating these currents for use in their high power station at Nauen.

The method of using these alternators is to connect them directly into the aerial circuit, which is tuned to have a natural frequency equal to that of the alternator. By means of a suitable switch the duration of these high frequency currents is controlled so as to produce the necessary "longs" and "shorts" for telegraphic signalling.

In the Poulsen arc system an electric arc, similar to that in the ordinary electric arc lamp used for illuminating purposes, is maintained between copper and carbon electrodes surrounded by an atmosphere of methylated spirits or other hydrocarbon vapour. When subjected to a powerful magnetic field and suitably adjusted, this arc has the property of causing continuous oscillatory currents to flow in any oscillatory circuit connected across its terminals. These oscillations are of a frequency equal to the natural frequency of the oscillatory circuit, but appear to be slightly less regular than the oscillations produced by a high frequency alternator. If the oscillatory circuit is a suitably tuned aerial, continuous oscillations will be induced in it just as (or almost just as) when a high frequency alternator is used.

In these systems absolutely undamped wave trains are sent out from the aerial and the maximum of selectivity is obtained. At a receiving station, working

with a transmitting station sending out these undamped waves, special methods have to be adopted
to receive the signals, which will be explained later.
Intercommunication between one of these stations
and an ordinary spark station is not therefore possible
under ordinary circumstances although they will
interfere with one another freely.

A further advantage is that with a continuous
wave system the aerial is actively radiating during
the whole of a "dot" or "dash." To illustrate this
point with figures; take a spark system using an
oscillatory frequency of 100,000 cycles per second,
each train consisting of about 50 effective cycles, and
let the spark frequency be 500 per second. The
intervals between the sparks will then be $\frac{1}{500}$th of
a second. Of this $\frac{1}{500}$th of a second, the aerial will
only be actively radiating for $50 \times \frac{1}{100000} = \frac{1}{2000}$th of
a second. For the remaining $\frac{3}{2000}$ths it will be in-
effective. Had a continuous wave system been in
use the aerial would have been radiating for the
whole $\frac{1}{500}$th of a second and for the same range an
aerial of only about $\frac{1}{4}$ of the size would have been
necessary. As the aerial represents about one half
of the total cost of a modern station, this is a strong
point in favour of the continuous wave systems. There
are unfortunately practical disadvantages. The alternators are expensive, require very careful attention,
and have yet to prove their durability. With the

Poulsen arc difficulty is experienced in keeping it ready to oscillate when required and for as long as is required ; moreover it is inefficient, only a small percentage of the total energy supplied being radiated.

CHAPTER VI

THE RECEIVING INSTRUMENTS

THE detectors used for receiving signals are very sensitive devices enabling the small currents in a receiving aerial to be observed. They are used indirectly, being made to operate other instruments by means of which the presence of the current is made known either as a sound signal or by the deflection of a beam of light. The former method is the one now most generally in use, the sounds being produced by a pair of ordinary telephone receivers. By their means every series of oscillating currents is made to produce a corresponding series of sounds which enable the message to be interpreted. If the detector is arranged to deflect a beam of light, the message can be recorded photographically, which has the advantage of reducing the possibility of mistakes but is not nearly so sensitive as aural reception with telephones, and messages are liable to be rendered quite unreadable by strong atmospheric disturbances.

It may be pointed out that telephone receivers alone cannot be used because in the first place the inductance of the windings would allow only a very minute oscillating current to flow through them, secondly the diaphragm could not move appreciably at the frequency of the oscillating currents generally used, and lastly even if it could move at this rate it would give out a note beyond the range of audibility, for few ears can detect sound waves of a frequency above 25,000 cycles per second.

In the old days the metallic filings coherer was the detector universally used, but at its best it was not very sensitive compared with the modern detectors and suffered from the further disadvantage of being easily put out of action by vibration. It was quickly replaced by the far more sensitive and reliable magnetic, thermal and rectifying detectors now in use.

Marconi's Magnetic Detector is one of these and consists of a band of soft iron kept moving beneath the poles of a pair of horse-shoe magnets (see fig. 13). Just beneath these magnets the wire is made to pass through two small coils, the outer of which is connected to the telephone receivers and the inner one to the aerial so that the small oscillatory currents will pass through it. Every train of oscillatory currents then causes a "click" in the telephone. These "clicks" combine to form a distinctive sound which is interrupted in the same way as the sequence of the trains of oscillations. Thus the "dots" and "dashes" are reproduced in the telephones exactly as they are sent out from the transmitting station.

The winding of the coil through which the oscillatory current flows may be of quite low electrical resistance and consequently the instrument can be

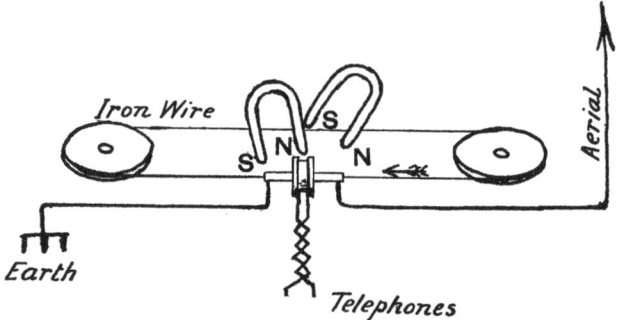

Fig. 13.   Marconi's Magnetic Detector

connected directly in the aerial circuit without introducing too much damping. This, however, is undesirable when a very selective circuit is required, as a single circuit, even of very low resistance, does not allow interference to be "tuned out." A better method, which is the usual procedure of the Marconi Co. with this detector, is to connect it to the aerial through intermediate resonant circuits, inductively coupled with one another. These circuits are very selective because they all have to be set oscillating in turn by impulses of the frequency to which they are all adjusted. Waves of other frequencies may set the aerial oscillating to some extent and the intermediate circuits very slightly, but unless the interference is very powerful the final detector circuit will be unaffected.

This detector is very stable and reliable, and the only attention it wants is regularly to wind up the clockwork which keeps the iron band moving. Much of the success of the Marconi ships' installations must be put down to its excellent qualities.

Fessenden's thermal detector consists of a very short length of exceedingly fine platinum wire sealed into an exhausted glass bulb, rather like the bulb of a tiny electric lamp. The oscillatory current is passed through this wire, heats it, and consequently causes an increase of its electrical resistance. This change of resistance is made use of indirectly to operate a telephone receiver.

A fundamental difference between the magnetic detector and this thermal detector may be pointed out. The heating of the wire in the latter depends on both the magnitude of the current and the time for which it is flowing. The same result can be obtained by a large current flowing for a short time or by a small current flowing for a long time. With the

magnetic detector, on the other hand, all the indica-
tions point to the fact that it is upon the maximum
value of the current that its action depends. A long
continued small current will not affect it to the same
extent as a large current flowing for a short time.
The thermal detector belongs to a class known as
integrating detectors, i.e. those which will add together
a series of small effects. With the use of more and
more sustained trains of oscillations, to secure greater
selectivity, detectors of the type of the magnetic
detector tend to fall into disuse, detectors of the
integrating type taking their place provided they
can be made equally stable and reliable.

The modern rectifying detectors of course belong
to the integrating class. They depend for their action
upon their property of being in some way or another
able to store up a little electricity from each train of
the oscillatory currents in the aerial. This charge
they give out again as a pulsating unidirectional
current to telephone receivers, one pulsation corresponding to every train of oscillations. One method
of connecting them up to the receiving aerial is shewn
in fig. 14. The aerial is connected to earth through
the inductance L_1. A second inductance L_2 and a
condenser K_1 form a second oscillatory circuit tuned
to the same natural frequency as the aerial and so
placed that the two circuits are inductively coupled
together. The detector D is connected across the

terminals of the condenser K_1 in series with another
condenser K_2.   The telephone receivers T through
which the pulsating currents are to be passed are
connected across the terminals of this condenser K_2,
and for the best results with most detectors a battery
B is connected in series with them.

The action of this apparatus is as follows.   When

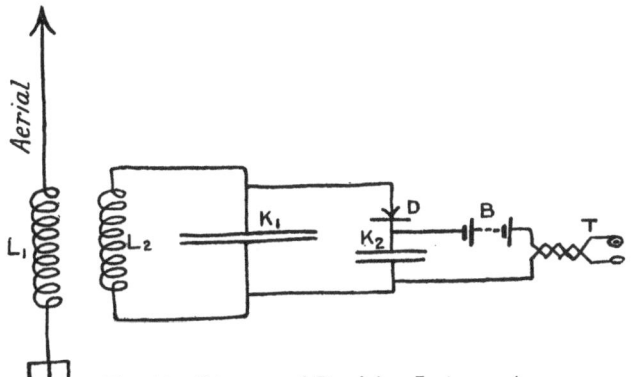

Fig. 14.   Diagram of Receiving Instruments

a train of waves of the right frequency strikes the
aerial, oscillatory currents are set up in L_1 which
induce oscillatory electromotive forces in L_2.   As
the two circuits are tuned to the same frequency,
these oscillatory electromotive forces will set up large
currents and there will be large oscillatory differences
of pressure between the terminals of the condenser

K_1. These of course tend to make currents flow
through the detector into or out of the condenser
K_2. The detector is so adjusted, however, that these
currents can flow one way but not the other, with the
result that the condenser K_2 becomes charged. For
suppose that the current can flow downwards through
D and not upwards. Then when the upper terminal
of K_1 is positive a charge passes through the detector
and collects on the upper part of K_2. When the
pressure is reversed the tendency is for this charge
to run out again. This the detector prevents. Thus
every time that the upper terminal of K_1 is positive
the charge in K_2 will be added to, and by the
end of the train quite an appreciable charge will
have been deposited in K_2. As it is collecting,
this charge will have started flowing comparatively
slowly through the telephone receivers. It would
appear at first sight that it would run out in this way
as fast as it collected. It does not do this on account
of the inductance of the receiver windings. For every
train of oscillations, therefore, the condenser K_2 has
a charge given to it in a series of rapid jerks which
it gives out again to the receivers as a single
pulsation of current in one direction. Each current
impulse causes a "click" in the receivers and if
continued rhythmically gives rise to a clear musical
note. The function of the battery in series with the
telephones is to maintain a steady current through

the detector and the telephone by way of the inductance L_2. The detectors requiring this battery are
found to be better rectifiers when they already have
a certain steady current flowing through them as well
as the oscillations. The adjustment of this current
is generally rather critical. The oscillatory currents
and the impulsive telephone currents are super-
imposed on this steady current, and the ultimate
result, viz. a regular variation of the current in the
receivers, is the same. Owing to the aerial and
intermediate circuits being tuned together this arrangement of receiving instruments is fairly selective.
Only powerful interference of a different frequency
from that of the signal being received will set up
sufficiently large currents to prevent the message
being read.

Many of these rectifying detectors consist of
crystals or crystalline materials which have this
property of unidirectional conductivity when placed
in contact with one another. In some cases, car-
borundum for instance, the property appears to lie
in the crystal itself; in others it seems to be at the
point of contact. This latter is the case with such
combinations as galena and graphite or a copper
point on iron pyrites. Some of them are more sensitive with the battery in series with the telephones
but others work better without it. There seems to be
no guiding principle in this and it is a matter of trial

to find whether a battery is an advantage or not. The disadvantage of these detectors is that most of them require very careful adjustment and are easily rendered insensitive by vibration or heavy atmospheric discharges.

Another detector coming into this class is the Fleming oscillation valve, which depends for its action on the fact that if a piece of platinum or other metal is sealed into the globe of an ordinary incandescent electric lamp, then, even although this metal makes no electrical contact whatever with the glowing filament, a current can be made to flow from it to the filament with a comparatively small voltage but not in the opposite direction. The Marconi Co. use this detector with both carbon and tungsten filament lamps, and in America a detector called the Audion is used by de Forest which, if not identically similar, is at any rate on the same lines. Both types are fairly sensitive when correctly adjusted and are not influenced by vibration or heavy discharges. They both require a battery in the telephone circuit, the pressure from which must be carefully adjusted. The current flowing through the filament must also be regulated, there being usually two or three sensitive points.

The electrolytic detector may probably also be included in the class of rectifying detectors. It has been made in various forms, usually consisting of a

fine platinum point immersed in a weak acid solution
in a lead container. If a battery of suitable electromotive force is connected up with one terminal to the
platinum point and the other to the container, a
transient current flows when the circuit is first closed,
which quickly dies away owing to the platinum point
in some way insulating itself from the solution. If
now an oscillatory pressure is applied between the
point and the container as well as the steady pressure
of the battery, a rectifying action takes place the
exact nature of which is uncertain and has led to
long discussion in scientific circles. This detector is
used with the same arrangement of receiving instruments as the other rectifying detectors, and whatever
its action may be it is remarkably sensitive when the
battery pressure is adjusted to the critical value,
being able to detect currents which would be quite
unnoticed on the magnetic detector. Like the crystalline detectors it is liable to be rendered insensitive
by heavy atmospheric discharges.

The receiving arrangements for the undamped
waves sent out by transmitting stations using high frequency alternators or oscillating arcs are necessarily
different from those just described. If for instance
an ordinary rectifying detector were used, the result
would simply be an increase of the current through
the telephone during each signal and a "click" at the
beginning and end of it only. It would obviously be

very difficult to pick out the "dots" and "dashes" under these circumstances. One of the best methods of receiving these undamped waves is that due to Poulsen. The aerial is connected to earth through an inductance, with which a second very efficient oscillatory circuit carefully tuned to the same natural frequency is inductively coupled. By means of a little instrument known as a "ticker" the condenser of this circuit is disconnected from the circuit and connected to a pair of telephone receivers at regular intervals. Whilst it is connected to the oscillatory circuit larger and larger oscillatory currents accumulate in it, gradually reaching quite high values on account of the very slight damping of the circuit. At whatever instant the condenser happens to be disconnected from this circuit it will be charged to a greater or less extent. It might happen on very rare occasions to be disconnected when it was not charged at all, but the tendency is for the break to take place at the instant of zero current, i.e. at the instant when the charge in the condenser is a maximum. When the condenser is connected to the telephone receivers this charge runs out and causes a "click." These "clicks" will take place every time the condenser connections are changed over if the waves are still coming in. Thus if the waves are interrupted to indicate "dots" and "dashes" the sounds given by the telephone will be similarly broken up. It will be noticed that the note heard in the telephone is in this case dependent not on the transmitting station but on the speed at which the ticker is working. This speed is generally so arranged that the ticker changes the connections of the condenser at the rate of about one thousand times per second. Reception by this method is very sensitive and selective but is unfortunately not as immune from atmospheric interference as would be expected.

CHAPTER VII

THE USES OF WIRELESS TELEGRAPHY ON BOARD SHIP

WIRELESS telegraphy serves two distinct purposes on board ship, firstly for safety and use in emergency, and secondly for general communication between ship and ship and between ship and shore. Taking the former of these, it is estimated that not less than 5000 persons have been rescued from serious predicaments through its agency, the majority of whom would in all probability have otherwise lost their lives. Numerous cases could be quoted. The collision between the s.s. "Florida" and s.s. "Republic," when the operator of the latter ship achieved world-wide notoriety for the manner in which he stuck to his work under the most adverse circumstances, is a case in point. More recently the P. and O. liner "Delhi" went ashore on a practically desert coast of Northern Africa. By the aid of wireless, assistance, which could not otherwise have been obtained, was forthcoming in a few hours. Or again, what the sufferings of the survivors of the "Titanic" disaster would have been but for their timely rescue by the "Carpathia," it is impossible to say. The inhabitants of the distant island of St Kilda will also appreciate the value of wireless, as it was by its aid that a cruiser was despatched to them with food supplies in the spring of 1912, within a few hours of their famished condition being known.

The utility of wireless in this connection is not only after a disaster has taken place. It now serves as an invaluable means of circulating information with respect to dangers to navigation in the form of ice, derelicts and bad weather. The tracking of typhoons in the Indian Ocean and China Sea has now been reduced to a fine art, and all ships in these seas receive regular warnings and are able to adjust their courses so as to evade the danger. Further, experiments are now being carried out with several types of so-called "Wireless Compasses," which enable a ship to find the direction from which a signal is being received. Thus on a dangerous coast, if a lighthouse or lightship is fitted with wireless apparatus, it becomes "visible" to passing ships fitted with these compasses even in the densest fogs or the worst snowstorms. The French Government are fitting up several stations for this purpose only, designed to give out regular intermittent signals so that they may be identified in the well-known way in which lighthouses are recognised at night, by the regular interruption of their beams of light.

To a smaller extent, the sending of "time signals" to ships far out from land is of value, because unless a navigator knows the Greenwich Mean Time accurately it is difficult for him to fix his position on the ocean with any degree of certainty. The "time signals" are therefore intended to enable him to check the accuracy of his chronometers. The signals are sent out at prearranged times by the powerful shore stations, e.g. the German station at Norddeich gives the signal at 12 midday, and the Eiffel Tower at 10 a.m. and 12 midnight; and by prearranged signals, one definite part of which is the hour to be indicated. The method used by Norddeich is to give some three or four minutes' warning so that operators may be ready, and then to give four series of five "dots," the last of which is 12 o'clock. This method allows very accurate checks to be taken, owing to the rhythmical manner in which the series of "dots" are sent out.

Wireless is thus a great addition to the safeguards of navigation as well as being invaluable in case of accident, and it is for both these reasons that many countries, e.g. Great Britain, The United States of America, Spain, Italy, etc., are making or have already made laws to the effect that ships beyond a certain size and going beyond a certain distance shall be equipped with reasonably good apparatus for the purpose. At the 1912 International Conference it was unanimously decided to recommend all Governments to make it compulsory for all ships, cargo or passenger, to be fitted with wireless instruments and to have on board at least one fully qualified operator. The fact that in Lloyds' Register of Shipping a special section has for some years been devoted to those ships fitted with wireless, shews the importance attached to it by ship-owners and underwriters.

The arrangement for distress signals to take precedence over all others under all circumstances has been in existence from the first. It was one of the earliest provisions of the Marconi organisation, the distress signal then being "CQD." At the 1906 International Conference this was changed to "SOS," but the Marconi operators still used the "CQD" signal among themselves, and it is interesting to notice that the first distress signal sent out from the "Titanic" was "CQD." After distress signals urgent Government messages, English or otherwise, take second place in the order of precedence, being followed, before any private or press messages whatever, by messages relating to navigation such as warnings and weather reports.

For general communication on board ship, wireless is used for much the same purposes as the ordinary telegraph on land. On all the frequented routes, a ship fitted with wireless apparatus is never isolated from the world. If it is not in direct communication with a shore station itself, it will be within range of some other ship which can take any desired messages and "relay" them on to a shore station. In this way it is now possible to start from, say London, and travel completely round the world without ever being out of telegraphic communication with the starting point. Across the Pacific from Japan to Vancouver and between New Zealand and San Francisco are perhaps the few regular routes on which uninterrupted communication cannot be depended upon during both day and night. Even these possible chances of being isolated from the world will shortly be removed. On the North Atlantic routes, every liner has the necessary apparatus and can be communicated with from either Poldhu or Cape Cod for the whole passage. This has made it possible to publish small daily newspapers, "The Atlantic Daily News" and the "Cunard Bulletin," on board these ships, giving all the most important news of the day, market fluctuations and reports, stock exchange quotations, and so forth. The busy business man or financier is thus no longer out of touch with the world for five or six days every time he crosses the Atlantic. Before the introduction of wireless, these crossings were looked upon as mental rests with the mind free from all worry and trouble. Those days are now past.

At most post and telegraph offices in Europe and America, messages are now received for transmission to ships within range of the ordinary shore stations at rates very little exceeding those of an inland telegram. In Great Britain and America messages are accepted for any ship on the Transatlantic service, to be sent at extra cost *via* Poldhu or Cape Cod if the ship is out of range of the ordinary shore stations. On other frequented routes the arrangements made are on similar lines, but necessarily not so complete. In time, however, it is probable that there will be the same continuous communication for a ship in any part of the world, whether it is on a regular route or not.

The value of this continuous communication is not to the passengers only. To the shipping companies alone, wireless more than justifies its existence. The exact time of arrival of a ship is now known beforehand, together with a full knowledge of the docking accommodation needed, the stores required for the next voyage, and any information with respect to repairs which may be necessary. By the time the ship arrives everything is in readiness, and the time saved is generally nearly a day, and often more, per voyage. The record Christmas trip of the "Lusitania" in 1911 from Liverpool to New York and back, was only possible under these conditions.

To the submarine cable companies, wireless has been in one way a veritable godsend. All their repair ships are now fitted, and instead of having to come to a port after every repair, they now get their instructions by wireless and can proceed direct to wherever they are next required, often saving voyages of several hundred miles.

In the fishing fleets in the North Sea and in American waters, the larger boats are now being fitted. The value of communication between ship and shore, whereby the supply of fish reaching the markets is known beforehand, can easily be understood.

Wireless has not infrequently been the means whereby escaping criminals have had their flight summarily cut short. The sensational discovery and arrest of Dr Crippen and Miss le Neve on board the s.s. "Montrose" is a case in point which will be well remembered.

At the present time, with the exception of some 30 or 40 vessels, all the wireless ships are under the control of one or other of the Marconi companies, and working under the Marconi organisation with the shore stations of the different countries. These companies train and to a large extent control their own operators, and have introduced a regular service similar to that of the Cable Companies' services. The operators have definite qualifications imposed upon them by the International Convention and by the Governments of the countries to which the ships belong, but beyond this their control is in the hands of the Marconi companies subject to the commanding officer of each ship whilst actually at sea. The operators sign on at the beginning of each voyage, and are given the relative rank of a junior officer. At the beginning of each voyage, sailing charts are prepared from which the operator can tell every day and every hour what ships should be within range. The list of these is generally posted on the ship's notice boards at intervals for the information of the passengers. Every ship as soon as it is within range of another, promptly calls it up and sends off to it any telegrams which may have been accumulating for it during the previous few hours.

In the larger and more important liners two operators are carried, and a continuous watch is kept day and night. During a voyage across the Atlantic these operators have their hands fairly full, sending and receiving telegrams to and from shore stations and other ships. Some three to four hours of the night are devoted solely to the reception of press news from Poldhu or Cape Cod for insertion in the following morning's newspaper. Ship to ship communication ceases during this interval, except for distress signals or very urgent messages.

In the smaller ships only one operator is carried, with the result that the watch is not continuous. This is a disadvantage as distress signals may be missed. The 1912 Conference has recommended that in ships of this class, the operator or other competent person capable of reading Morse should listen for the first ten minutes of every hour, and that distress signals should always be made at those times as well as at others. By this means a ship near another one in distress would be certain of picking up the distress signals within an hour.

Although such a large proportion of the ships have their wireless under the control of the Marconi companies, they are by no means all of them fitted with Marconi apparatus. All the systems used can, however, intercommunicate freely so that no difficulties are experienced. The arrangement of the apparatus in its main features is very much the same in all cases. It of course depends upon the size of the ship, as the requirements of an Atlantic liner are very different from those of a trawler. In the one case skilled operators are available, and in order to get long ranges and rapid signalling more complicated apparatus can be employed than in the other, where the operator has other duties and only just

sufficient wireless training to be able to send and
receive messages in Morse.

The power of the machines supplied for transmitting purposes varies from about one-half horse-
power up to 10, giving effective ranges of from 50
up to 400 or 500 miles when working with the medium
power shore stations.   The aerials consist usually of
from two to six horizontal wires arranged lengthways
of the ship and suspended from the highest possible
points of the masts.   One or two connecting wires
are brought down from these elevated ones to the
wireless cabin, which is generally situated on one of
the upper decks not far from the bridge, with which
easy communication is provided. These wireless cabins
on the larger ships are divided into three parts, namely,
sleeping quarters for the operators, a space for the
transmitting instruments, and the operating room
with the receiving instruments. Figs. 15 and 16
shew the transmitting compartment and the operating
room of a ship fitted by the Marconi Company. The
lower part of the transmitting compartment contains
a "rotary converter," which is a machine for converting the usual direct current from the ship's
electrical supply, to an alternating one of fairly high
frequency. This alternating current is taken to a
transformer placed on the shelf above, which raises
the voltage to that required for charging the condensers placed on the same shelf alongside it. These

Fig. 15. Arrangement of instruments in the transmitting
compartment of a Marconi ship

condensers are made up of glass sheets with layers of metallic foil between them, the whole being immersed in insulating oil. The discharge circuit of the condensers is through the primary inductance, contained in the box fixed on the right-hand side high up on the back of the compartment, and the spark gap placed on the shelf above the rotary converter and marked "2," fig. 15. In the latest type of transmitting instruments, a rotary spark gap on the shaft of the rotary converter is used. The secondary inductance is split up into two parts, one of which is in a box in front of the primary inductance with which it is inductively coupled, and the other is in a second box attached to the back of the compartment on the left. The wires from the aerial are connected to the latter, one end of the former being connected to the earth connection which is attached to the steel frame of the ship itself. These two circuits thus constitute the two circuits of the ordinary coupled system. Both primary and secondary inductance can be varied to enable several tunes to be used. The usual tunes worked on are the commercial 300 and 600 metre waves, but the provision is there for others if they are required. In the operating room will be observed, on the right, a switchboard and apparatus for the control of the rotary converter in the transmitting compartment. On the kneehole table, to the right, is an ordinary

Fig. 16.   The operating room in a Marconi ship

induction coil which can be used for transmitting
distress signals in emergency if the supply of electricity from the ship's dynamo fails. To the left of
this coil is the "multiple tuner," a device for rendering
the receiving circuits very selective. Above it is the
magnetic detector and to the left of it Fleming valve
detectors, either of which may be used at will. The
use of these two types of detector is a luxury provided
on the larger ships only, the smaller ones having the
magnetic detector alone. A point of interest in the
practical use of the multiple tuner is a small "change-
over" switch by means of which it may be put into
or cut out of the receiving circuits. It is cut out
when the operator is keeping a "general lookout."
His receiving circuit is not then very selective, and
any signals over quite a wide range of wave-length
may be picked up. This is an advantage because all
ships nominally on the same tune may not have exactly the same adjustments, so that a very selective
receiving circuit adjusted to one ship might miss
signals from another. With the non-selective receiving circuit this is almost impossible. When the
operator hears his ship called, he then puts in his
tuner and adjusts it so as to get the best results for
the particular message.

Fig. 17 shews one type of transmitting apparatus fitted by the Telefunken Company, except for
the motor-alternator which is placed in a separate

Fig. 17. Telefunken Transmitting Instruments

compartment. This latter consists of an alternator giving alternating currents of a frequency of 500 cycles per second driven by an electric motor. The current from this alternator is taken to a transformer shewn under the table on the left-hand side. Here the voltage is increased to that suitable for charging the condensers, which can be seen in the form of four vertical tubes at the back of the table. These condensers consist essentially of glass tubes coated inside and out with thin metallic foil. The inside layers are all joined together and form one plate of the condenser, the outside layers the other, the intervening glass serving as the dielectric. The discharge circuit of these condensers consists of the Telefunken multiple spark gap, the cooling vanes of which can be seen over the back of the table on the right, and the primary inductance under the table, not visible in the figure. This inductance is adjusted by pushing a plug into suitable holes on the table. The aerial is connected up to the inductances on the left of the table on the white porcelain insulators, thence to the primary inductance and so to the earth connection, a direct or conductively coupled circuit thus being used. Two oscillatory circuits thus "conductively" coupled have exactly the same properties as the two "inductively" coupled circuits of the Marconi apparatus. Adjustments of the inductances in the aerial and primary circuits are provided for tuning purposes.

Fig. 18. Telefunken Receiving Instruments

Fig. 18 shews the receiving apparatus. By means
of the switch at the back of the instrument, the aerial
is changed over from transmitting to receiving. At
the top of the instrument are two inductively coupled
coils. One of these is connected to the aerial and
to earth, the other to moving vane air condensers
seen in front of the apparatus. These two circuits
are adjusted for fine tuning by turning the handle
of the air condenser. Coarse tuning is obtained
by replacing one or other of the coils by the spare
ones shewn at the right-hand side. The detectors
are arranged to slip into holes which can be seen
just behind the condensers, the electrical contact
being automatically made by the two projecting
clips. The detectors are thus easily changed if they
become insensitive. The telephones have moveable
plug terminals which fit into sockets at the side
of the detectors. The adjustment of this receiving
circuit is very simple, and yet at the same time it
is very fairly selective. Electrolytic and crystalline
detectors of a very permanent and sensitive type
are used. In some ships fitted with this apparatus
tuned microphones are used, tuned to the spark
frequency of the transmitting station, to amplify the
sound produced in the telephones. This, however,
is at the expense of considerable complication of the
adjustments which are easily thrown out by vibration.
Also, as would hardly be expected, atmospherics and

other interference appear to be magnified to the same
extent as the signals which are being received.

CHAPTER VIII

THE SHORE STATIONS

By this term is generally meant those stations
situated on or near the coasts for the particular
purpose of communicating with passing ships, and
which are open for intercommunication with all ships
on one or other of the commercial wave-lengths.

In Great Britain there is a complete set of these
stations round the coast which are under the direct
control of the General Post Office.   In Germany,
Italy, France, and most other countries, a similar
control is exercised, but in America there are shore
stations belonging to private firms as well as to the
Government, both of which are open for commercial
use.   A few countries, e.g. Germany, Italy and
Holland, have individual stations of much longer
range than the majority of the English stations.
The geographical fact of the British Isles lying right
on the more important of the great ocean highways
renders a large number of stations necessary, but
owing to the existence of the powerful Marconi
stations it has not so far been necessary for the

Post Office to establish long range stations of their own. For other countries, however, if they wish to communicate with their ships directly as soon as they are within range of land they will generally have to do so over long ranges. The majority of the shore stations are connected up by land lines to the telegraphic systems of their own countries. A few of them, situated on isolated islands are really cable stations on which the wireless installation acts as a feeder to the cable. A good example of a station of this kind is the one in the Cocos Islands in the Pacific. The Eastern Extension Australia and China Telegraph Companies' cables pass this island, and the wireless instruments belong to the company and are worked by their staff.

The duties of these stations lie almost entirely with the transmission of messages from ship to shore and *vice versa*, which by the 1906 Convention it is their duty to regulate when the traffic is heavy. As well as this primary function they also act as efficient coast-guards, in that the distress signals of any ship going ashore would be certainly picked up by them and assistance sent.

In Great Britain the shore stations are Crook-haven in the extreme S.W. corner of Ireland, the first station with which the Atlantic liners are in touch and consequently a very busy one requiring a staff of six operators working day and night in pairs ; Rosslare, a station used by shipping passing to Liverpool; Seaforth, near Liverpool, for the use of steamers entering that port ; The Lizard, another busy station, being the first to come into touch with the ships coming up the English Channel; Bolt Head, the first station taken over by the Post Office, and Niton in the Isle of Wight near St Catherine's Point, both of which are used in connection with the up-channel traffic; the North Foreland for the use of ships entering or leaving the Port of London and passing through the Straits of Dover ; Caistor in Norfolk for use over the southern part of the North Sea ; Cullercoats, near Newcastle, for service with ships on the N.E. coast and on the North Sea; and lastly Malin Head in the north of Ireland for service with ships from the Clyde and Liverpool passing to the northward of Ireland.

As well as these Post Office stations there are a certain number of more or less private ones used on particular services only. For instance, the London, Brighton and South Coast Railway have such stations at Newhaven and Dieppe for use with their cross-channel steamers between those ports, and the Great Western Railway have a station near Fishguard for use with their Irish boats.

On the Continent, Germany has a complete set of stations along both the Baltic and North Sea coasts. At Norddeich on the extreme western point of

Friesland there is a powerful station capable of communicating with ships well out into the Atlantic. In Holland is another powerful station used for the same purpose, at Scheveningen, near The Hague. France maintains stations along her northern coast as well as in the Mediterranean. In Spain a very complete network of powerful stations has recently been opened. There is a centrally situated station at Aruanjuez near Madrid with which the shore stations proper at Vigo, Cadiz, Teneriffe, Las Palmas, Soller (in the Balearic Islands) and Barcelona, maintain regular communication. Portugal is following suit in the near future with stations connecting up Lisbon with the Azores, Madeira and St Vincent. Ships on the South American and South African routes will thus have a number of stations with which to communicate on both outward and homeward journeys. Italy of course has a very complete system round the whole of her coastline, and there are several stations in the Eastern Mediterranean. In Canada and America the chain is complete round both eastern and western coasts and on the Great Lakes. Many of the principal islands of the West Indies have stations, and there are a few scattered round both the Atlantic and Pacific coasts of South America. In the Pacific there is an important station in Honolulu, and three in the Fiji Islands used partly for communication with passing ships and partly for ordinary telegraphic purposes between themselves. Australia and New Zealand have rather lagged behind for various reasons, but are now making up for lost time. Round the coasts of Asia stations are sparsely distributed from Petropavlovsk in Eastern Siberia to Aden in the West. Japan has a few stations, and there are several in the Dutch East Indies. In Africa a number of stations have been erected in the Italian East African colonies; there is one at Durban, another at Cape Town, and there are a few on the western coast.

CHAPTER IX

THE USE OF WIRELESS TELEGRAPHY BETWEEN FIXED STATIONS OVER LAND AND SEA

For the purposes of signalling between fixed and permanent stations, whether over land or sea or both, wireless comes immediately into competition with the older telegraphy by submarine cable and land lines.

In comparing the advantages of the two systems in any given case four principal points have to be dealt with, namely, Reliability, Rapidity of Transmission, Cost, and Secrecy.

Reliability has been placed first and is without

doubt the most important adjunct of any telegraph
service.   At present the older system has probably
a slight advantage for both short and long distances.
Wireless signals may be delayed or wrongly transmitted for three principal reasons, namely interference
or atmospherics, personal errors in sending or receiving,
and failure in the instruments.   The two last possi-
bilities are common to both the old and new systems
equally, for the personal element is the same for both,
and against the greater possibility of failure of a
wireless station must be placed the chance of damage
to the cable or land line.   Against the difficulty of
atmospherics and interference must be placed the
possibility of magnetic storms which entirely dis-
organise the ordinary telegraphy, but have no effect
on wireless.   There is a case on record in which
Mr Marconi failed to get a message through by cable
owing to one of these storms, and had to fall back
upon a partially finished wireless station, with
successful results.   Magnetic storms of this magnitude are, however, much less frequent than atmospherics, and for this reason cable or land wire
telegraphy must be said to have the advantage.
Methods of eliminating the effects of all but the very
worst of these wireless pests are however proceeding
so rapidly that there is every hope of the new system
becoming as reliable as the old one in the course of
a few years.

With regard to Rapidity of Transmission wireless
is at present by no means as far advanced as the
ordinary telegraphy, which has had something like
50 years' start and has reached a high state of
perfection, whereas rapid transmission by wireless
is only just developing.   With cable telegraphy it is
possible to arrange the instruments so that messages
can be sent both ways between two. stations at the
same time, a method known as "duplex working."
The same results can be obtained with wireless.

The question of Cost is intimately connected with
the question of the speed of working.   Obviously the
more work that can be got out of a system in a given
time the cheaper will be the service.   Given equal
rates of working the relative cost of the two systems
must depend entirely on the conditions.   Land wire
or cable will have to be used in some cases, quite independent of cost, e.g. for telegraphic communication
between London and Birmingham or Manchester.
Leaving out cases of this kind and dealing with those
in which both systems are possible, the choice between
the two will ultimately depend upon the cost of construction and maintenance of a cable or land wire as
compared to the cost of erection and maintenance
of the wireless stations.   Generally speaking the
constructional cost is less for the wireless stations.
With regard to maintenance the advantage will lie
with wireless in tropical and uncivilised countries,

where continual clearing and patrolling is necessary for the land lines or in places where the shores are rocky and the cost of maintenance of cables is high. It must of course be remembered that wireless signalling over long distances requires a good deal of power, and the running costs of the power station must be considered in comparison with the maintenance costs of land lines or cables. The wireless station has one advantage, namely that if there is a breakdown it is on the spot with a staff and appliances for repair immediately at hand, whereas with a land line or submarine cable if there is a failure it has first to be located and then repaired, and a long time may elapse before communication is re-established.

Finally in comparing the two systems there is the question of Secrecy to be considered, one to which a good deal of importance has been attached. It is an undoubted fact that wireless under all conditions is not as secret as cable or land-line telegraphy. But this objection is probably far less serious than appears at first sight. It is presumably in connection with commercial work that the difficulty is supposed to arise as the wave-lengths are fixed and universally used. Consequently if, say, a ship is sending some message to a distant shore station, all surrounding ships and perhaps several other shore stations will be able to read the message. But it must be remembered that it is only the operators at these stations who read

it, and they are under the same obligations as to the
secrecy of any message they overhear as they, or the
cable operators, are to any message which they
actually send or receive.   Private stations may of
course also receive the message, but they are all
known and licensed, and in Great Britain, at any
rate, one of the conditions of granting the licence
is that no use shall be made of any messages which
may be picked up.   It is unlikely that any leakage
could go on for any length of time without being
detected and promptly put a stop to.   In the United
States, where up to 1912 little or no control of private
stations was exercised, the difficulty is (or was)
greater, but from recent events it would appear that
the trouble there was more in the direction of false
messages being sent out than in legitimate ones being
read and the information thus gained improperly used.
As a last resort some system of code can always be
made use of, as well for wireless as for other telegraphy.   This proceeding is of course only applicable
to short messages.   Press news could hardly be
transmitted in that way owing to the labour of
coding and decoding.   In many cases, for example
Marconi's transatlantic service, this kind of news is
rendered sufficiently secret by the fact that it is not
worth anyone's while to put up a sufficiently large
aerial to "tap" the message privately.   The news
can of course be picked up on a small aerial near to

the station from which it is being sent, but it will usually be of no value there. Nine-tenths of the press news from Poldhu could be picked up on a small aerial anywhere in the South of England, but it would be of no value to anyone after they had got it.

Summing up, it may be said that for some time to come there is a wide and useful field for both systems in which they will actually assist one another, but that in the future competition between wireless and submarine cable telegraphy is inevitable over both long and short distances. Similar competition will also in all probability arise between wireless and long land lines.

At present Marconi's transatlantic service is more advanced than any other case of telegraphy between fixed land stations by wireless. A brief history of the development of this service, together with some indication of the difficulties which had to be overcome before the now regular communication could be maintained is given in Chap. XIII. Both at Clifden and at Glace Bay the stations are in wild and isolated positions overlooking the Atlantic. A large amount of power is required, and consequently a large and self-contained power installation is necessary at each station. The wireless part of the apparatus is similar on both sides. Separate sending and receiving aerials are used to radiate long waves so as to enable the

large distances to be covered even under the worst
conditions.   To enable a large amount of power to
be used and for these long waves to be efficiently
radiated enormous aerials are necessary.   At Clifden
the transmitting aerial is about 170 feet in height
and extends from the power house for over three-
quarters of a mile in a direction straight away from
Glace Bay.   The earth connections are taken to a
lake in a direction directly towards Glace Bay.
With this arrangement the signals are transmitted
powerfully in a direction towards Glace Bay, but
comparatively weakly in other directions.   The
receiving aerial is of the same shape, but still longer,
extending for over a mile horizontally.   Another
peculiar feature of these stations is that high voltage
direct current is used for transmitting purposes,
obtained from several specially constructed dynamos
all connected in series and working in conjunction
with a battery of about 6000 accumulator cells.   The
condensers are also rather unusual, consisting of
large zinc sheets suspended from heavy insulators.
Alternate sheets are joined together forming the
plates of a large parallel-plate condenser having air
for its dielectric.   This is rather a cumbrous arrangement, but has the advantage that if by accident it is
over-charged and sparks across between the plates,
the dielectric is self-healing.   With a glass-plate condenser such an occurrence necessitates the removal

of the damaged plate which may take some time.
A rotary spark gap is used so as to ensure regular
sparking. The high voltage direct current supply
is connected straight up to the condensers, charging
them up in the intervals between the sparks; and to
prevent a large current rushing from the dynamos
and battery when the spark takes place, inductively
wound coils are also connected in the circuit from
the dynamos. Thus every time the projecting studs
of the wheel come opposite the fixed ones a spark
takes place and the condensers discharge with
powerful oscillations. Inductively coupled with the
primary inductance is a secondary one connected to
the aerial and earth with an additional inductance
in series for tuning purposes. The arrangement is
therefore just the same as the usual coupled transmitting circuit with the high voltage alternating
supply replaced by the direct. The "dots" and
"dashes" are made by means of a magnetically
operated switch in the circuit to the dynamos. This
switch required very careful design. Electrical
engineers will appreciate the difficulty of breaking
from two to three hundred horse-power at a pressure
of nearly 20,000 volts at the rate of about 200 times
a minute. This duty the switch has to perform day
in and day out with unfailing regularity. For receiving purposes Mr Marconi says that he uses the
Fleming oscillation valve with complete success, no

doubt with some highly selective form of receiving circuit so as to prevent interference from other signals as well as from atmospherics.

A fairly regular public service has been maintained since 1908, one of the difficulties experienced being, curiously, with the connections to the land telegraph systems, especially on the Canadian side, where they are exposed to bad weather conditions. Arrangements with the Postal Authorities are now complete, and telegrams can be handed in at any Post Office for transmission from England to America and *vice versa*.   The Transatlantic traffic however is so large that the Cable Companies have not at present felt this competition to any appreciable extent.

With one exception the Clifden and Glace Bay stations are the most powerful in existence.   This exception is a very large and imposing station erected by the Marconi Co. in Italy at Coltano in the Gulf of Genoa.   Details of this station have not been published, but it is understood to be designed for a power of 1000 horse-power, and is to be used for communication with the Italian East African colonies, with Italian ships on the South American and South African routes, and possibly with the South American stations themselves.

Other well-known high-power stations are the Eiffel Tower station in Paris, the large experimental

station at Nauen near Berlin, belonging to the Telefunken Co., and the American stations at Arlington
near Washington and at Brant Rock, on the eastern
coast, some 20 miles south of Boston. The Eiffel Tower
station is under the control of the French Government
War Departments and is used largely for experimental purposes. Few details of the instruments in
use have been published. The aerial is in the form
of an inverted fan suspended from near the top of
the tower and spreading out over the surrounding
gardens at an angle of about 45 degrees. It is the
highest aerial in existence at present, and some very
long ranges have been obtained with the use of
comparatively small powers. The transmitting and
receiving instruments are placed in an underground
compartment partly for secrecy and partly so as not
to disfigure the gardens more than is necessary. The
latest installation is reported to be of 225 horse-power,
using a spark frequency of 1000 per second.

Nauen is principally interesting on account of its
aerial. This is of what is termed the "umbrella"
type, and is suspended from one large mast. The
original mast (see fig. 19) consisted of two parts with
a ball and socket joint at the bottom and at the
junction between the two and was 650 feet in height.
The lower part was of heavy steel lattice work construction of triangular cross-section, the sides of
which were 13 feet in width. The upper part was

of similar construction but much lighter, and both parts were supported by strong steel guys anchored to heavy masses of concrete.   This very bold construction failed on account of one of these guys breaking in a heavy gale with the result that the whole mast and aerial collapsed.   The present mast is of similar construction but about 900 feet in height, making it very nearly equal in height to the Eiffel Tower.   The aerial wires are spread out in all directions from the top of the mast to a number of small masts placed on the circumference of a circle of about 500 yards in diameter, some of which can be seen in the background.

The large American stations were erected partly for experimental purposes and partly for communication with the naval ship and shore stations.   They are both situated on high land and have large umbrella-type aerials suspended from masts some 600 feet in height.   Power for transmitting purposes is obtained from alternators giving a high spark frequency and using rotary spark gaps on the alternator shaft.   It is proposed to use very high frequency alternators and undamped waves.

The part played by wireless in the development of South America, principally in the Amazon region, is one of its most interesting applications to the duties of ordinary inland telegraphy.   Up to 1906 the only regular communication over the whole of this vast

Fig. 19.   The Nauen Wireless Station, near Berlin

area consisted of a cable from Para to Manaos and a very leisurely postal service carried by the steamers up and down the river. This region is particularly one for wireless as opposed to land line telegraphy. For a land line wide clearances would have to have been made and maintained through the forest, the poles would have needed protection against both insects and animals and constant patrolling and inspection would have been necessary. The cable it is true could have been extended, but this would have been expensive, and interruptions of the existing cable were frequent. Both Marconi and Telefunken stations are in use and there is now a complete line of communication from Para, at the mouth of the Amazon, to Lima on the Pacific coast of Peru. Stations have also been erected to the south of this line in Peru, Brazil and Bolivia, forming a system of about 30 stations in all connecting up all the most important settlements. The crossing of the Andes between Lima and Equitos was a new departure in wireless engineering successfully carried out by the Telefunken Co. In the Congo a similar development has taken place. Here the difficulties in the way of land wires are even greater than in South America. No telegraph poles have yet been designed capable of standing up against a herd of wild elephants.

Lastly there is the great Imperial Wireless Scheme, not yet certainly an accomplished fact because it has yet to receive the sanction of the House of Commons. This scheme was entered upon after full discussion at the Imperial Conference in 1911, partly for strategical and partly for commercial reasons. It is felt that a limited number of wireless stations can be more effectively protected than many thousand miles of cable which in case of international difficulties could be so easily cut. The proposed arrangements between the Postmaster-General, acting for both Home and Colonial Governments, and the Marconi Co. as to the cost and maintenance have been made known, but technical details have only been published in part. Six stations are suggested, situated in England, Egypt, the East African Protectorate, the Union of South Africa, India and near Singapore. In addition to these another powerful station at Fort Darwin, Australia, is proposed, which is to be erected by the Australian Government but which is to form a part of the whole scheme. These stations will each have to be able to communicate under all conditions with the stations on either side in the chain, and the one in East Africa with three others, viz. South Africa, India and Egypt. The stations are to work duplex at 20 words per minute or with a simplex automatic system at 50 words per minute. The Postmaster-General will pay the Company £60,000 per station in two sums, £40,000 when the stations are erected and the remaining £20,000 after they have

been satisfactorily worked by the Company for six
months. For 28 years after the completion of the
scheme the Postmaster-General will pay the Company
10 per cent. of the gross receipts by way of a royalty
in return for which they have the free use of all the
Marconi patents as well as the technical advice of the
Company. The wave-lengths to be used are from
17,000 to 50,000 feet. Care is to be taken to prevent
interference between the stations included in the
scheme as well as existing large power stations near
the route. Large aerials of course will have to be used
and as much advantage of directional effect will be
taken as possible. Power will be provided by alternators driven by steam turbines varying from 1300 to
2500 horse-power. The Marconi type disc discharger
directly coupled to the alternator shaft is to be used
and a different note will be given to each station to
assist in the elimination of interference. The transmitting condensers will be of the parallel-plate type
immersed in oil and the usual coupled transmitting
circuits will be employed. At the terminal stations
the steam generating plant and the turbo-alternators
are to be duplicated, one set of transmitting instruments only being proposed. At the intermediate
stations a complete set of power generators and
transmitting instruments is to be provided for each
station with which communication will have to be
maintained, together with one set of spare boilers

and turbo-alternators. Thus in the East African station there will be three complete sets of transmitting instruments, including primary condensers, primary and secondary inductances, and three aerials. This multiplication of the apparatus is necessary in order that communication may be maintained with all three surrounding stations at the same time, a state of affairs that may easily arise at times of heavy traffic. The 2500 horse-power turbo-alternators, of which there will be four sets, will be installed in this station and when working simultaneously in all three directions 7500 horse-power will thus be in use. The Clifden and Glace Bay stations pale into insignificance compared to this. It should be noticed, however, that the power of all stations is expected to be in excess of that actually required. In consequence of the enormous amount of power to be handled the signalling switches for breaking up the signals into "dots" and "dashes" are in triplicate throughout, with means of changing over from one to the other very quickly in case of failure. Signalling at 50 words per minute, the circuit to the primary condensers will have to be made and broken at the rate of about eight times a second, and the power to be broken may, at this switch, be about 2000 horse-power. It would not be surprising if these switches presented greater difficulties both in design and operation than any other part of the scheme.

The receiving stations are to be situated not less than ten miles from the transmitting stations, this distance being necessary for duplex working. No details of the proposed receiving instruments are published at present.

The contract for this scheme came in for very adverse criticism on its publication, and a Select Committee of the House of Commons was appointed to consider it. At the time of writing only an interim report has been made by the Committee, who are awaiting the results of an enquiry by an independent Scientific Committee into the relative advantages of the different systems.

Several other ambitious schemes on similar lines are also mooted. For instance, the Marconi Company are said to have started with the erection of a chain of large stations connecting up England with N. America, California, the Sandwich Islands, and Japan. The French Government has been reported to have in hand a scheme connecting up Europe with S. Africa, S. America, and the Pacific. Germany, with the aid of the new Nauen aerial, proposes to maintain communication with N. America and her colony of Togo in central Africa. If all these schemes materialise the tuning will have to be very accurate and very selective circuits will be necessary if serious interference is to be prevented.

CHAPTER X

THE USES OF WIRELESS TELEGRAPHY FOR NAVAL AND MILITARY PURPOSES

EFFICIENT means of communication are one of the most important essentials of modern warfare and it is not surprising that both naval and military authorities have carefully watched the development of wireless from the very first. Before its introduction a fleet at sea was dependent for its information upon high speed cruisers. Signals could not be transmitted beyond the limits of vision either by day or night, and in fog communication between ships was possible by sound signals only. All this isolation has now gone, with far-reaching results. An army in the field has for years been able to signal over long distances by means of temporary land wires. These wires, however, were easily cut and consequently required constant protection in an unfriendly country. Wireless is not subject to this disadvantage as both scientific training and the necessary apparatus would be required to interrupt the service. Thus for Military uses wireless has been merely an advance on the old system but for Naval purposes it has been a revolution.

It should be remembered that wireless signals can be received by an Enemy as well as a Friend, so that war-time may perhaps be the quietest of times from a wireless point of view. A destroyer prowling by night near an enemy's fleet would be as likely to use her wireless as switch on her searchlights. Her presence would be given away as much by one as by the other. Unless used with the greatest care therefore wireless may be a source of danger, and it is possible that the results of the whole of a series of operations may depend upon the skill with which it is employed. On account of this the great American strategist, Admiral Mahan, has gone so far as to predict that in the near future it will be safer in some circumstances to fall back upon the old system of sending information by fast cruisers than to use wireless.

Wireless has been used in actual warfare on several important occasions. At the time of the Boxer trouble in China temporary stations were fitted up at the Taku forts and were found invaluable for communication between them and the ships outside. In the Russo-Japanese war it played a prominent part, all the principal ships on both sides being fitted, and the Russian army having at least three portable sets in use. The Japanese both outside Port Arthur and before the battle of Tsushima were able to keep their principal ships at a safe distance and yet remain in constant touch with their cruisers and destroyers who were watching the Russians. The latter were as a matter of fact on some occasions warned of the intentions of the Japanese by overhearing and deciphering their signals.

It is curious that there is no record of the Russian ships in Port Arthur having made any use of their wireless to maintain communication with the outside world.   Presumably the distances were too great for the early type of apparatus with which they were fitted.   Could regular communication have been kept up even during the first six months of the siege it is possible that the ultimate end of the war might have been quite different from that which was actually the case.   The Russian general, Stössel, would have been compelled to hand over supreme control of the defence to the Fort Commandant, General Smirnoff, and other inefficient officers would have been superseded, with the probable result that at the very least, the capture of the fortress and the destruction of the fleet sheltering in the harbour would have been delayed for several months.   This would have greatly added to the difficulties of the Japanese as it might have allowed the Russian Baltic Fleet to effect a junction with the Port Arthur ships, in which case the Japanese fleet would have been inferior in numbers and the results of a battle would have been doubtful.   Had the Japanese been defeated at sea they could not have continued the campaign on land and victory would have rested with Russia.

During this war a small steamer, the "Hainum," was fitted with de Forest apparatus and used with a similar shore station near Wei-hai-wei for the purpose of collecting first hand information for the *Times* newspaper. Some annoyance was caused to both belligerents and the steamer was probably very lucky in escaping from being sunk or at any rate from being seized and having her wireless gear removed by one or other of them.

The Italians took a number of Marconi portable sets to Tripoli, but the extent to which they were used is rather uncertain. The large station at Derna, which had previously been used for communication with Turkey, was destroyed early in the war but was refitted by the Italians for their own use between their ships and their army.

Little is known at present of the extent to which wireless has been used in the Balkan War. Adrianople has certainly been in touch with the Turkish head-quarters during practically the whole of the siege, but beyond that practically no mention of its use has been made. That a number of portable sets of instruments have been available on both sides is known, and if ever the veil of secrecy is lifted it is more than probable that wireless will be found to have played no unimportant part.

CHAPTER XI

WIRELESS TELEGRAPHY ON AIRSHIPS AND AEROPLANES

THE fitting of wireless apparatus to airships and aeroplanes has generally been in connection with their military duties.   The advantage of wireless communication to an airship or aeroplane when engaged in scouting operations is obvious.   It will become to them in the near future what it now is to a warship, but at present this application, especially in the case of aeroplanes, is in quite an experimental state.   The great difficulty lies in the fact that no earth connection is possible.   To radiate or absorb any electromagnetic energy it is therefore necessary to arrange for two sets of conductors to form a radiative circuit similar to that used by Hertz.   With airships the difficulty is overcome by using the body of the ship as one conductor and a long trailing wire for the other, thus converting them into a huge vertical Hertz oscillator.   This arrangement would radiate waves which could be picked up on the usual type of earthed aerial if the two were tuned to the same frequency.   The same device has been tried with aeroplanes but with little success.   On one occasion a Farman biplane was fitted with two parallel trailing

wires which together made up the radiative circuit.
These wires were naturally found to be a nuisance on
the aeroplane to say nothing of the danger of their
becoming involved with the propeller, and the latest
arrangement is to attach two sets of wires to the

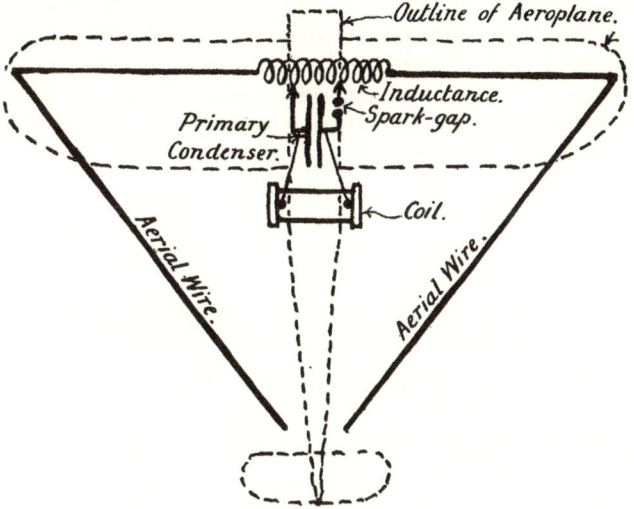

Fig. 20. Arrangement of Wireless Apparatus fitted to Aeroplanes

planes themselves as shewn in the diagram. The
capacity of one of these sets with respect to the
other will obviously be very small on an aeroplane
whose greatest dimension does not much exceed
20 feet, and consequently waves of only about 300

feet in length can be efficiently used. The transmitting instruments must obviously be as light as
they can possibly be made. On airships comparatively
powerful sets can be employed, absorbing some two or
three horse-power which can be obtained from the
main engines, but on aeroplanes small batteries have
to be used, the greatest output from which does not
exceed about one-tenth of a horse-power. So far as
details have been published it appears that one form
or other of the usual coupled transmitting circuits has
been used. The arrangements shewn in the diagram
are those fitted to an aeroplane used in trials carried
out on Salisbury Plain in connection with the Annual
Manœuvres of 1910. Signals from such an arrangement could not be exchanged with an ordinary earthed
aerial because the waves given out are in the wrong
plane. Some form of horizontal radiator and absorber
was necessary and took the form of two similar sets
of wires arranged fanwise in a horizontal plane, the
apices of the two fans being near one another, with
the receiving and transmitting instruments between
them.

In comparison with the wireless apparatus in use
on board ship and in the land stations these arrangements appear crude, and it is not surprising that no
very long ranges have been attained. Receiving
signals appears to be more difficult than sending
them, and there are at present only one or two

records of them having been received at all on
aeroplanes. In the case of the Wellman airship on
the other hand the receiving was better than the
sending, probably owing to the small amount of
power used for the latter purpose.

The German military airships are all stated to
have been fitted, and have been at various times
reported to have maintained communication over
long distances. The Clement-Bayard airship which
came over from Paris to London in October 1910
was equipped but the extent to which the apparatus
was used is uncertain. The ill-fated Wellman airship
" America " had a small Marconi set of instruments
and a Marconi operator on board which were found
useful, though curiously when the airship had to be
abandoned no reply could be got to the distress
signals, and the attention of a passing steamer was
called during the night by hand signalling with a
lamp. The steamer presumably carried only one
operator who was not then looking out for signals.
The large French air cruiser "Adjudant Vincénnot"
was able in September 1911 to maintain easy communication with both the Eiffel Tower station and
with one at the fortress of Verdun during the whole
of a 13-hour flight.

Turning to the aeroplanes, during the experiments
on Salisbury Plain distances up to three miles were
obtained, and better results have since been reported

with similar apparatus. In France the Farman biplane with the two trailing wires was able to transmit signals to the enormous Eiffel Tower aerial over distances up to 40 miles. In May 1911 a German aeroplane was reported to have been in touch with the large station at Nauen during the whole of a flight from Berlin to Hamburg. Very recently after a long flight from Eastchurch to Portsmouth, Commander Samson is said to have reported his safe arrival to officers at the former place by means of the wireless apparatus fitted to his hydro-aeroplane.

CHAPTER XII

WIRELESS TELEPHONY

ALMOST from the first introduction of the telephone, efforts have been made to get rid of the connecting wires. A long series of experiments were made in Germany by Ruhmer with a " speaking arc " and a selenium cell, but the apparatus was cumbersome, inefficient, and of no practical use. The "conductive-inductive" method of telegraphy between two parallel wires was used for telephony with better results. A microphone transmitter was connected in series with the sending wire, and when spoken into, the sounds were reproduced in the telephone receivers in the receiving wire. This method was used by Preece between the Skerries and Cemlyn and also between the island of Rathlin and the mainland with such success that the former of these two installations is still in use. The range of this method is, however, limited, even when the parallel wires can be made several miles in length.

The only system of wireless telephony really presenting any hope of being of use over long distances, is the one using the same electromagnetic waves as wireless telegraphy. Fessenden and de Forest in America have devoted a great deal of attention to this system since 1899. The former was in 1908 successful in transmitting speech quite clearly from Brant Rock to New York, a distance of about 400 miles. In the same year, de Forest apparatus was fitted to several of the United States warships making their famous round-the-world tour. The installations do not seem to have been of much practical use although no difficulties were experienced in communicating over distances up to 50 miles. In Germany the Poulsen system was adapted to telephonic transmission with some success and ranges of 250 miles were obtained. A number of systems, de Forest, Poulsen and others, were tried at the Eiffel Tower with the result that speech was clearly transmitted on one occasion over a distance of 155 miles. During 1909 the French cruiser "Condé" was fitted, and distances up to 100 miles reached. The British Admiralty were also at this time reported to have been carrying out trials with the de Forest apparatus, but with only moderate success. Latterly it appears that rather less interest has been taken in the problem, at any rate fewer "inventors of the wireless telephone" have appeared.

To understand the action of the Wireless Telephone it will be necessary to touch very briefly on the question of the vibrations of the air set up when a person is speaking. The act of speaking consists in expelling air from the throat and mouth in such a way that certain particular vibrations are set up corresponding to the particular sounds it is desired to produce. These vibrations are of small amplitude and high frequency and are of course quite independent of the bodily movement of the mass of the air. Some of these vibrations are quite regular ones of definite frequency, as for instance those which go to make the sound ōō in "Coo." Others such as ē in "Me" or ā in "Ma" are a little more complex, being made up of combinations of two or three simple vibrations of different frequencies. The consonant sounds, on the other hand, are much less regular and do not consist of vibrations which repeat themselves periodically. The problem of the transmitter with all telephony is to make use of these air vibrations to produce changes in an electric current which shall be exact reproductions of the air vibrations. If this can be done the currents can be used to reproduce the sounds at a distance in a telephone receiver. For the ordinary telephony with wires the air vibrations are thrown up against a thin metallic diaphragm by means of the trumpet mouthpiece, causing small movements in it exactly similar to the movements of the air. Behind this diaphragm and between it and a fixed plate is a small quantity of fine granulated carbon. A battery is connected up with one terminal to the diaphragm and one to the fixed plate. When the diaphragm is set vibrating the contact resistance of the granules varies in a manner nearly proportional to the magnitude of the diaphragm's motion. This variation of the resistance causes a variation of the current from the battery, which can be made use of to cause corresponding movements in the receiver diaphragm. The latter diaphragm in turn sets up vibrations in the air surrounding it. If the whole motions have been faithfully reproduced throughout, then these vibrations given out will be of the same nature as those taken in, and to an ear placed near the receiver the sounds will appear the same as those produced at the transmitter end.

With wireless telephony practically the same process has to be gone through. In the transmitting aerial continuous oscillations have to be maintained whose magnitude can be varied in exact reproduction of the vibrations constituting the sounds which have to be transmitted. Variable contact devices similar to those used for ordinary telephony are as a matter of fact employed. The receiving aerial, which is tuned electrically to the same frequency as the transmitting aerial, thus receives a continuous train of waves of varying magnitude. With a suitable detector connected up with telephone receivers in exactly the same way as for wireless telegraphy, these variations of the magnitude of the waves can be converted into variations of the magnitude of the current in the telephone receivers. Suppose for example the sound ōō is being transmitted. Before the sound begins the transmitting aerial is giving out, and the receiving aerial is taking in, a continuous wave of invarying amplitude. At the receiving end this is accompanied by a steady current through the telephone receivers and no sounds are produced. But after the speech has started there will be regular rhythmical variations of the intensity of the waves given out and received. With increase of the intensity the detector in the receiving aerial rectifies more of the oscillations, and with decrease of the intensity, less. The result therefore is that the current in the telephone receivers increases and decreases in the same way as the intensity of the waves, and the sounds produced in these receivers will be the same as those actuating the transmitter at the transmitting station. For the consonant sounds the same process goes on except that the variations are irregular.

It may be noted that the electrical wave frequency has no direct connection with the frequency of the air vibrations. Electrical wave frequencies of from 50,000 to 1,000,000 are used, whereas the frequency of the air waves varies from about 200 to 5000. It is obvious that the frequency of the electrical oscillations must be much greater than that of the air vibrations, because for each variation of the latter a series of electrical oscillations is required from which current for the telephone receivers can be rectified. If this was not the case each variation of the air movements would not be reproduced and the speech would not be clear.

Continuous oscillations in the transmitting aerial are not absolutely essential for wireless telephony. Spark systems may be used as long as the spark frequency is not less than 10,000 per second. As each spark is accompanied by a train of oscillations there will be currents in the aerial practically the whole of the time and the behaviour is very much the same as with continuous oscillations.

For satisfactory telephonic communication there are three principal requirements; firstly the source of continuous, or practically continuous, oscillations ; secondly a means of varying the intensity of these

oscillations in the transmitting aerial corresponding
to the air vibrations; and thirdly a detector at the
receiving end whose rectifying action is proportional
to the intensity of the incoming waves. The first and
third of these requirements present no serious difficulties. The continuous oscillations of the required
frequency are produced by high frequency alternators
of the Alexanderson type as used by Fessenden in
his Brant Rock-New York experiments or by the
Poulsen or de Forest oscillating arcs. For the
detectors, the electrolytic and several other integrating detectors have the necessary properties. The
second requirement, however, is the difficulty which
has yet to be overcome. For telegraphic signalling
over ranges up to, say, 500 miles some two or three
horse-power is required in the aerial. The variations of this power are from the full power during the
signals to zero in the intervals between the signals.
To telephone over the same distance with the same
aerial nearly as large a variation of power would be
necessary. But these variations of power have to
take place, not at the rate of three or four times per
second, but at rates of from 200 to 5000 times per
second, and the only source of energy that is available
to effect this variation is that which can be obtained
from the vibrations set up by the voice. Looked at
from this point of view the results which have actually
been obtained are almost unbelievable. Microphones

similar to those used for ordinary telephony are generally employed, but with modifications to enable them to deal effectually with the large currents and pressures which are necessary in the transmitting aerials. Water cooling devices have been tried with these microphones to make their action more regular, as it was found that overheating was one of the reasons for the carbon granules setting together and being unaffected by the vibrations of the diaphragm. In some cases several microphones have been used in series or in parallel with improved results. Various devices, too, have been tried intended to relieve the microphone of its very arduous duties by arranging it merely to throw the transmitting aerial out of tune instead of directly varying the magnitude of the current in it. As a matter of fact the work the microphone has to do is not so much reduced as appears to be the case, and the improvement is not very great. The difficulty, in fact, seems at present to be insuperable, and is the barrier in the way of the extensive use of the wireless telephone.

The utility of wireless telephony if these difficulties could be overcome lies in the fact that it would be more rapid in its action and would not require a practised operator to read the signals. Anyone understanding the language in use would be able to communicate by its means, and arrangements could be made for two or three people to carry on a conversation together. In any case its application would of course be limited to ships and possibly to a very few long distance installations. Its use in a busy town with several thousand subscribers is inconceivable.

CHAPTER XIII

HISTORY

THE history of Wireless Telegraphy, or to give it its official title, Radio-telegraphy, must be regarded as dating from the publication of Maxwell's famous paper, "A Dynamical Theory of the Electromagnetic Field," read before the Royal Society on December 8th, 1864. It is true that signalling between stations by electrical means without the use of connecting wires had been carried out earlier than this, but the methods employed differ from those of modern wireless and, although they continued to be used until as late as 1897, their practical application is very limited. In his paper Maxwell proved theoretically the existence of electromagnetic waves, but the practical confirmation of the theory was not forthcoming until 1878 when Hertz published the results of his experiments with oscillating currents. Following this many eminent physicists took up the subject and the possibility of using the waves for signalling purposes occurred to some of them. It is interesting to note that in some of Hertz's experiments tuned circuits were used, and that in 1894 Lodge had demonstrated before a large audience the possibility of signalling from one room to another through several walls.

It remained however to Marconi to produce the first set of apparatus specifically constructed for signalling from point to point by the agency of electromagnetic waves. He brought his apparatus, which was of the simple type now known as "plain aerial," to England early in 1896 and took out his first patent in June of that year. Marconi's great advance lay in the use of the earth connection which is one of the most important points covered by this first patent. Many demonstrations were given in the presence of representatives of the various Government Departments and other prominent people, and as more experience was gained greater ranges were obtained. By 1899 signalling across the English Channel had been accomplished and the East Goodwin Lightship had been placed in communication with the South Foreland Lighthouse. This is perhaps the first practical application of wireless. A few ships and shore stations had been equipped and the difficulties of interference immediately presented themselves. To overcome them, Lodge, in 1898, patented his method of using more sustained oscillations by means of which more advantage could be taken of the tuning of the transmitting and receiving stations. For the same purpose Marconi in 1900 introduced the coupled transmitting circuit with fixed spark balls, the famous "four sevens" patent being taken out in that year. Others besides Marconi and Lodge were working on the subject, notably Braun and Slaby in Germany, Fessenden and de Forest in America, each contributing his share to the general advance and in many cases taking out patents for similar devices at almost the same time. These many patents led later to long patent actions in which, with the exception of the Lodge patent, Marconi has been consistently successful, from which it must be assumed that he was, on the whole, ahead of his competitors.

The coupled circuit system with improvements is the system in almost universal use to-day. The tendency of the changes has been in the direction of increased efficiency, higher spark frequency and longer wave-lengths. The spark gap has probably been subject to the greatest alterations. The fixed spark balls being found unsatisfactory for high spark frequency, Marconi gradually adopted a rotary gap. At first this was driven by a separate small motor, but during 1910, and after, it was fitted to an extension of the alternator shaft. A similar arrangement had been successfully used by Fessenden in 1906 in a station in Scotland used for signalling across the

Atlantic. In the same year, 1906, Wien, in Germany, pointed out the advantages of "charging by impact." This was taken up by the Gesellschaft für Drahtlose Telegraphie, the manufacturers of the "Telefunken" apparatus, who use it in their now well-known "singing spark" instruments introduced in 1909.

The practical use of continuous oscillations also dates from about 1906 when Poulsen developed his oscillating arc apparatus into a more or less reliable system. Some three or four years later means of generating these continuous oscillations directly from special types of alternators were devised by Alexanderson in America, and Goldschmidt in Germany. Still later, in 1912, a third method was introduced by the Telefunken Company.

The development of the receiving instruments has been similarly progressive. The metallic filings coherer of 1896 proved unreliable and was quickly superseded. In America and Germany thermal and electrolytic detectors were used. Marconi's magnetic detector was introduced in 1901 and immediately proved itself superior for general use to any other detecting instrument then known. The modern rectifying detectors and their highly selective circuits came in by degrees from about 1906 onwards.

Marconi's early efforts at signalling across the Atlantic are not without interest. The first attempt was made in December, 1901. The Poldhu station,

the largest then in existence and having installed in
it machinery of some 30 to 40 horse-power, was just
completed. A temporary aerial attached to a kite
was sent up at St John's, Newfoundland, and the prearranged signals from the Cornwall station were
picked up on three consecutive days. Having thus
proved the possibility of signalling over this distance,
large stations were designed for Cape Cod and Cape
Breton, and an increase in the size of the aerial
was arranged for at Poldhu. This work was completed in 1902 and signalling in both directions was
carried out. For the regular communication required
for commercial purposes larger aerials and more
power were soon found necessary. The large Clifden
(Ireland) and Glace Bay (Nova Scotia) stations were
taken in hand in 1906, and after many preliminary
experiments a fairly regular communication was
established on commercial lines in 1908.

Soon after Marconi's first demonstrations in
England a company, "The Wireless Telegraph and
Signal Co., Ltd." was formed to work his patents and
manufacture his apparatus. This company, the name
of which was changed in 1900 to "Marconi's Wireless
Telegraph Co., Ltd.," occupies a very important place
in the history of wireless telegraphy. A related company, "The International Marine Communication Co.,
Ltd." soon followed, which took over the working of
the apparatus installed in the ship and shore stations,

the organization of the signalling and the control and training of the operators. Subsidiary manufacturing companies in which the original company is financially interested, have also been formed in America, Germany, France and many other countries. Owing to the decisions arrived at in recent patent actions, here and abroad, these companies have for all practical purposes a world-wide monopoly. Their only important competitors at present are the Telefunken and Poulsen Companies, and possibly, in the near future, the Compagnie Universelle de Télégraphie et de Téléphonie sans Fil, who own the Goldschmidt patents.

Early in its development the necessity for some international agreement with respect to wireless signalling became obvious. In 1903 a conference was held at Berlin without definite results. A second was held in 1906 and a long Convention was signed by the delegates. The principal provisions of this Convention were :

1. The fixing upon 300 and 600 metres as the standard wave-lengths for commercial purposes ; a range of from 600 to 1600 metres being reserved for naval and military uses.

2. An agreement that all stations should intercommunicate freely, independently of the type of apparatus with which they were fitted.

3. A number of service regulations relating to

methods of calling up and answering, distress signals, control of tariffs, etc., etc.

This Convention of course required ratification from the various Governments before becoming law in their countries. Many Governments ratified at once. The Marconi Co. looked upon the intercommunication clause as an attempt to rob them of the advantages which they were gaining from their private organization, their operators at that time not being allowed to communicate with stations fitted with other than Marconi apparatus. They therefore protested against ratification of the Convention both in England and America. In England a Parliamentary Committee was appointed to go into the question fully. After hearing a vast amount of expert evidence the Committee decided by a majority of one in favour of ratifying, which was consequently carried out. The United States, Italy and Japan did not ratify immediately, but by the time of the holding of the third Conference in London in 1912 they had all come into line. At this third Conference many details were arranged for the improvement of the service, but the most important matter dealt with was the question of the use of wireless for saving life at sea. This question was put before the Conference by the British delegates at the request of the Government as a direct consequence of the "Titanic" disaster. The result of the discussion was a recommendation that all ships above a certain size should be equipped with wireless apparatus and carry qualified operators.

At the same time that international regulations were being evolved, most countries found it necessary to exercise some control over the use of wireless in their own territories. In Great Britain the Wireless Telegraphy Act was passed in 1904 and is still in force. Its most important enactment is that no station of any kind may be used except under licence from the Postmaster-General. In this way the indiscriminate use of powerful transmitting apparatus by amateurs is largely prevented, whereas in America, where no such act was in force, the result was chaos.

BIBLIOGRAPHY

A Handbook of Wireless Telegraphy. Erskine Murray. (London. Lockwood and Sons.)

A History of Wireless Telegraphy. Fahie. (Edinburgh. Blackwood and Sons.)

A Manual of Wireless Telegraphy. Collins. (New York. Wiley and Sons.)

An Elementary Manual of Radio-Telegraphy and Radio-Telephony. Fleming. (London. Longmans and Co.)

Die Elektrische Wellentelegraphie. Arendt. (Brunswick. F. Vieweg und Sohn.)

Electric Waves. Hertz. English translation by D. E. Jones. (London. Macmillan and Co.)

Electromagnetische Schwingungen und Drahtlose Telegraphie. Zenneck. (Stuttgart. Enke and Co.)

La Télégraphie sans Fils. Poincaré. (Paris. Naud.)

La Télégraphie sans Fils et les Ondes Electriques. Boulanger et Férrié. (Paris. Berger-Levrault.)

La Télégraphie sans Fils. Van Dam. (Paris. Ch. Beranger.)

Leitfaden der Drahtlosen Telegraphie. Zenneck. (Stuttgart. Enke and Co.)

Les Oscillations Electriques. Poincaré. (Paris. Carre.)

Les Oscillations Electriques : Principes de la Télégraphie sans Fil. Tissot. (Paris. Doin et Fils.)

Principles of Wireless Telegraphy. Pierce. (New York. McGraw-Hill Book Co.)

Radio-Telegraphy. Monckton. (London. Constable and Co.)

Signalling across Space without Wires. Lodge. (London. The Electrician Publishing Co.)

The Principles of Electric Wave Telegraphy and Telephony. Fleming. (London. Longmans and Co.)

Wireless Telegraphy and Wireless Telephony. Kennelly. (New York. Moffat Yard and Co.)

INDEX

For EU product safety concerns, contact us at Calle de José Abascal, 56–1°,
28003 Madrid, Spain or eugpsr@cambridge.org.

www.ingramcontent.com/pod-product-compliance
Ingram Content Group UK Ltd.
Pitfield, Milton Keynes, MK11 3LW, UK
UKHW010851090126
466816UK00011B/154